Safety and Health for Workers - Research and Practical Perspective

Edited by Bankole Fasanya

Published in London, United Kingdom

IntechOpen

Supporting open minds since 2005

Safety and Health for Workers - Research and Practical Perspective
http://dx.doi.org/10.5772/intechopen.77424
Edited by Bankole Fasanya

Contributors
Igor Bello, Elvira Nikšić, Edin Beganović, Anna Maria Giammarioli, Helena Almeida, Bankole Kolawole Fasanya, Moruf Adegbite, Maged Milhail, George L. Stefanek, Alejandro Orgambídez Ramos, Carina Martinho Santos

First published in London, United Kingdom, 2020 by IntechOpen
IntechOpen is the global imprint of INTECHOPEN LIMITED, registered in England and Wales, registration number: 11086078, 7th floor, 10 Lower Thames Street, London, EC3R 6AF, United Kingdom
Printed in Croatia

British Library Cataloguing-in-Publication Data
A catalogue record for this book is available from the British Library

Additional hard and PDF copies can be obtained from orders@intechopen.com

Safety and Health for Workers - Research and Practical Perspective
Edited by Bankole Fasanya
p. cm.
Print ISBN 978-1-78984-524-2
Online ISBN 978-1-78984-525-9
eBook (PDF) ISBN 978-1-83881-043-6

For EU product safety concerns:
IN TECH d.o.o., Prolaz Marije Krucifikse Kozulić 3, 51000 Rijeka, Croatia, info@intechopen.com or visit our website at intechopen.com.

Meet the editor

Bankole K. Fasanya received a BSc in Mechanical Engineering in 1999 from The Polytechnic Ibadan, Nigeria, his Master's degree in Industrial and Systems Engineering from Morgan State University, Maryland, USA and his doctorate degree in Industrial and Systems Engineering specialized in ergonomics and human factors from North Carolina Agricultural and Technical State University, USA. His research focuses on human and environmental safety, ergonomics and human factors, auditory prevention and protection and noise assessment and control at workplaces. Dr. Fasanya is currently an assistant professor at Purdue University Northwest in Indiana, USA. He currently serves as one of the executive members of the American Hearing Conservative Association (NHCA). He is an OSHA-Authorized general industry safety train the trainer and a certified occupational hearing conservationist (COHC).

Contents

Preface

This book covers aspects of research and practical applications for safety and health issues encountered by individuals in various workplaces when performing daily tasks. Included are ergonomic hazards for women, student spine deformations, workers' health inequalities, social support for worker job satisfactions, good hygiene practices at the workplace, and practical applications to resolve issues. Detailed field findings and empirical studies are included in the book for better understanding for readers.

The intent of writing about safety and health of workers is to give those who are learning about human safety or workplace safety knowledge about what could negatively affect workers' wellbeing while performing their daily tasks. Unlike traditional articles, conference proceedings, and white papers that address only one topic each, this book presents a dual perspective on both research and practice. Likewise, there are many great books on safety and health management, several which are used as texts and references. Meanwhile, workers and individuals need to be knowledgeable in more than one source of workplace hazards. Additionally, in the health and safety issues at workplaces, workers need to be conversant not only with hazard controlling, but also to recognize the main causes of the hazards and part of the body predominantly bearing the consequences. *Safety and Health for Workers - Research and Practical Perspective* treats the subject from the point of view of the worker and individual who is interested in recognizing hazard sources, targeted body parts, and processes for controlling hazards. I have addressed those aspects that will be beneficial to the workers and individuals that may read this book for the purpose of preventing health and safety accidents.

Chapter I, "The Introductory Chapter" describes the general overview of the textbook and potential sources of workplace hazards that are constantly exposed while workers perform their daily activities. Included are ergonomics with real-life scenarios, hand hygiene, workplace conditions, the Workers Act, and products and raw materials.

Chapter II, "Women, Ergonomics and Repetitiveness" describes why females have lower biomechanical negative effects in the upper extremities compared to their male counterparts and a higher rate of productivity, especially in tasks of low force demand.

Chapter III, "Spinal Deformities with Students in Classroom Teaching in Urban and Rural Areas" describes the differences in the spine deformities of students in the urban and rural areas with regards to different postures. Discussion on the degree of spine deformity and the corrective measures is included.

Chapter IV, "Working Conditions and Health Inequalities" describes how working conditions and inappropriate gender inequalities in workplaces could negatively affect the implementation of good health and safety policies.

Chapter V, "Linking Social Support with Job Satisfaction: The Role of Global Empowerment in the Workplace" describes how social support in workplaces could affect the workers' job satisfaction. The mediator effect, structural empowerment, and social support in the workplace, especially in the healthcare industry, was reported.

Chapter VI, "Hand Hygiene Practices in Public Restrooms: Effects and Proposed Solutions" describes human common practices while using public restrooms, and the associated hazards and recommendations to improve hand hygiene practices.

Over the year, many people have contributed technically, intellectually, administratively, and morally toward the success of the first edition of this book. Importantly, I would like to thank Ms. Ivana who is the backbone of the success of this book, all chapter contributors (Prof. Bello Igor, Prof. Elvira Niksic, Edin Beganovic, Dr. Helena De Almeida, Dr. Giammarioli Anna Maria, Prof. George Stefanek, Dr. Maged Milkhail, Mr. Moruf Adegbite), Dr. Gail M. Augustine and Ms. Precious Fasanya for her valuable inspiration. Finally, I would like to thank my wife (Mrs. Olariyike Fasanya) and my son, Victor Fasanya for their unwavering support.

Bankole K. Fasanya
Purdue University Northwest,
Hammond, Indiana, USA

Chapter 1

Introductory Chapter: Safety and Health for Workers - Theory and Applications

Bankole K. Fasanya

1. Introduction

According to 2017 estimates of accidents and work-related diseases and deaths report released by the International Labor Organization (ILO), each year 2.78 million workers die from occupational accidents of which 2.4 million are disease related, approximately 86.3% of the total estimated deaths. The work-related mortality accounted for 5% of the global total deaths (based on the Global Burden of Disease Study 2015) and fatal accidents accounted for the remaining 13.7% [1]. Likewise, for 2017, BLS reported 282,750 MSD cases resulting in days away from work in the private sector, a continued decrease from the previous year (285,950) [2]. Work-related musculoskeletal disorders (WMSDs) accounted for 31.2% of all injuries and illnesses involving days away from work and remain the largest source of injury and illness cases [2]. Therefore, work-related issues have led to the explosion of research on safety and health for workers in the ergonomics and human factors, medical, psychological, engineering, science, nursing, and environmental literature for over two decades. The health and safety hazards address both acute and chronic hazards workers are exposed to at workplaces daily. These days health hazards at workplaces attract more attention because of the potential harm associated with the employee's exposure and the monetary cost attributed to a single cause. Thus, everyone wants a safe and healthy workplace for proper operation and productivity of the employees. Therefore, health and safety at work is aimed at creating conditions, capabilities, and habits that enable the worker and the organization to operate efficiently and in a way that avoids events that could cause them harm [3].

To reduce the rate of accidents reported every year at work, the United State (US) enacted a law to oversee the safety and health of workers at various workplaces in 1970, known as Occupational Safety and Health Act (OSHA). The act was enacted purposely to ensure that employers provide safe working environment for all employees. The importance of human safety and good health cannot be overemphasized or overestimated. Before the existence of OSHA, the National Environmental Protection Policy Act of 1969 became the basis of the environmental protection in the US. This chapter provides the overview of the book on safety and health for workers from research and practical application perspectives.

Everyone is exposed to at least one hazard daily and constantly interacts with different hazards on daily basis. The hazards concern with safety and health hazards, include biological, botanical, mechanical, chemical, physical, etc. Human beings cannot escape inhalation, egestion, injection, and absorption processes, even if they refuse to engage in daily work activities. Thus, it is imperative to study in depth the processes and the methods to control and manage the everyday hazards humans are constantly exposed to. Furthermore, to complicate issues, human typically operates under different constraints including inability to survive without interaction with one another, environments, materials, and equipment on daily basis to function as human. Given these constraints, human beings are forced to maintain good health while performing their daily tasks by using good judgment during decision-making. Exposure to hazardous materials is inevitable but human errors, time of exposure, concentration, and the dose of the hazardous materials determine the effect on safety and health of the workers. To reduce the likelihood of making errors, human beings must first understand the theory behind each operating process of any activity before engaging in it. Therefore, identification and recognition of hazards relating to the theory and the applications of the principles of handling the hazards are the significant ways for protecting human safety and health at work.

In a cafeteria in the northwest region of United State (US), a teenager rushed out from a restroom and everyone seated immediately perceived unpleasant odor. At once, everyone choruses what is this? The sister of the teenager boy asked her brother "did you wash your hands? The boy responded oh NO, I forgot." Immediately, debate on why and how started in the room that inspired other customers to join the discussion.

Likewise, Wickens et al. [4] document a situation in the Midwestern factory in US on an assembly-line worker. In the scenario, the assembly-line worker had to reach to an awkward location and position a heavy component for assembly. Toward the end of a shift, after grabbing the component, he felt a twinge of pain in his lower back. A trip to the doctor revealed that the worker had suffered a ruptured disc, and he missed several days of work. He filed a lawsuit against the company requiring physical action that endangered the lower back.

A young lady who graduated with GPA of 4.0 from one of the ivy league Universities in US, pick up a supervisory role position in a multi-million construction company and having worked for 5 years in this position, she realized that her health has been deteriorated. She visited the doctor and discovered she suffered from cardiovascular damage, gastrointestinal problems that has led as ulcers and irritable bowel syndrome, and decreased fertility. She requested for damage compensation from the company.

This book is designed to address the theory of safety and health for workers and the practical applications to solving issues associated with hand hygiene, social support, ergonomics-spine deformity, workplace conditions, health equality, etc. at workplaces.

2. What is ergonomics?

Many workers and employers constantly face one or more of the episodes described in the introduction. The episodes illustrate the role of ergonomics and human factors, industrial hygienist and safety specialist in workplace design. The umbrella for human safety and health at workplaces is known as the ergonomics and human factors. Various definitions exist for ergonomics and human factors from different sources such as scientific literature, professional organizations, governmental agencies, scholars etc. Ergonomics as defined by scientists is the science and practice of designing jobs and workplaces to match the capabilities and limitations of the human body. In 2018, it was precisely defined as fitting the job, workplaces, products to users' capability to enhance safety and good health of the users [5]. CETENA (The Italian Ship Research Company) defined ergonomics as the study of human performance and its application to the design of technological systems [6]. The author further stated that the goal of ergonomic is to enhance productivity, safety, convenience, and quality of life. It is a fact that ergonomics is derived from the Greek words ergo (work) and norms (law), which denotes the science of work. Therefore, the principles of ergonomics are not limited to traditional occupational environment as stated in Pamela McCauley-Bush textbook [7].

3. Hand hygiene

Hand hygiene is the act of cleaning hands for removing soil, dirt, and microorganisms. Proper hand hygiene had saved thousands of humans from death, chronic diseases, and incurable sickness. In the mid of eighteen century, Dr. Ignazz Philip Semmelwies proposed hand washing to reduce child mortality in the nation, before the discovery of microbial in the late eighteen century. At the time Dr. Semmlwies proposed the method, baby mortality was at 15–20%, and after implementing the proposed procedure, mortality rate reduced significantly to 0.85% in the hospitals where he worked while in other hospitals it remained at 10–15% of mortality rate [8]. Hand hygiene has been taken as the practice mostly needed for the medical practitioners. However, it should be considered necessary and important for everyone, as no one is exempted from sickness or diseases. Absolutely, people have forgotten that cleanliness is next to godliness as the saying goes. Therefore, complete hand hygiene is what is required in our workplaces today, even as many diseases spring up every day. Not long ago, Ebola disease took over the entire world, SARS also, and now it is Corona virus; so, who can tell what the disease will be tomorrow. Nearly, all the new outbreaks are traced to poor hygiene. Nevertheless, the importance of hand hygiene in the workplaces cannot be underemphasized therefore, the earlier the intervention, the better to preventing the spreading of any disease outbreak. In 2017, the findings of the study conducted by Prasad et al. [9] concluded that most patients and nurses interviewed reported hand hygiene as an important aspect of everyday life in preventing infection in the hospital setting. Many researchers have also suggested that hand hygiene is one of the main strategies to preventing infections that affect hundreds of millions of individuals worldwide and are constantly leading to chronic diseases, illnesses, and prolonged hospital stays, which sometime added burden to patients' financial stand and later cause instability within families [10–12]. There are many chemicals, substances, and materials human beings interact with on a regular basis at workplaces and homes, which have high potential of contaminating hands. Sometimes, people forget that their hands are dirty and pick food straight into their mouth without washing their hands, while many scratch their eyes 100 times daily and are found constantly touching their nose and seeping fingers in their mouth. Thus, educating people on the importance of hand hygiene is a crucial aspect of life to preventing unplanned sickness, diseases, and transfer of similar dangerous germs.

4. Workplace condition

Workplace conditions have been proved to have both positive and negative effects on employees' health status likewise on their productivity. The more the employee is satisfied with workplace conditions the faster the organization success, which makes workplace condition a major concern for organizations today. In 2019, Sorensen et al.'s study revealed that workplace condition significantly affects workers' health as well as their well-being [13]. Additionally, it is a proven fact that safe working conditions influence the habits of workers, which in turn impacts on efficiency [14]. Accidents at workplaces are mainly attributed to workplace condition or workers' acts. Workplace conditions refers to the working environment and all existing circumstances affecting labor in the workplace, including working time (hours of work, rest periods, and work schedules) to remuneration, legal rights and leaders' responsibilities as well as the physical conditions and mental demands that exist in the workplace. In other words, workplace condition deals with the organization leadership structure and floor layout or equipment setting. International Labor Organization (ILO) has consistently demanded for workers' work-life balance. Therefore, critical attention is required for worker schedule, leadership skills of the authority, and good workplace layout for proper flow to ensure workers' well-being and better productivity. Research findings on job redesign at call centers significantly improved employee well-being, psychological contract fulfillment, and supervisor-rated job performance [15], mental health [16]. Likewise, research has revealed that work systems design may have effects on physical health, mental health, and longevity of life [17]; therefore, continuous attention to health and safety of the employees is important because ill-health and injuries caused by the systems of work or working conditions cause suffering and loss to individuals and their dependents [18]. As simple as workplace condition might sounds, the fear created through improper workplace conditions such as violent, abuse, segmental judgment, gender biases, etc. are factors, which could prevent employee from taking goal-directed action toward safety. Although, the psychologically, flexible people are less emotionally disturbed in such environments as argued in Hayes et al.'s [19] findings. Nevertheless, adequate consideration should be given to workplace condition to ensure workers' well-being and safety at workplace.

5. Worker acts

Employees' act is another important aspect to be considered at workplaces, to reduce or eliminate accidents that might result in illnesses, diseases, injuries, and death at workplaces. Certainly, no environment is free from all hazards, but workers' behavior toward handling the hazards is a crucial aspect of preventing accidents from occurring at workplaces. Undoubtedly, safety and health specialists need to have general knowledge of the impacts of any hazardous materials to the human body and train all employees on ways of preventing such effects, but because of human nature, how many adhere to the totality of this principle? Employees' safety responsibilities were acknowledged in 1999 by Bratton and Gold [20] that employee is to behave in a manner that safeguards his or her own health and safety and that of his/her co-workers. Furthermore, Jonathan and Mbogo documented the findings from the research carried out at The Research Centre Design and Technology of the Saxion University of Applied sciences on "Safety at work," which acknowledged that personal safety, a safe environment, and safe behavior were important components that employers need to ensure their availability within their organizations for proper productivity and safety of the employees. Liao et al. [21] acknowledged in their study that almost all accidents and injuries occurring at workplace are attributed to workers' unsafe acts, which are also a reflection of system deficiency and unsafe work condition. Dodoo and Al-Samarraie [22] concluded in their study that lack of adequate knowledge on safety and health, violation of safety rules, work pressure, stress, and non-use of protective equipment were the unsafe acts perpetrated by the workers sampled. Researcher has also referred unsafe act as a deliberate omission or deviation from the recommended safety standards Mason [23]. The same researcher described an unsafe act as an individual's likelihood of not following standard safety rules, procedures, instructions, and specified criteria for work imposed by the organization management. Since the contributions of worker's unsafe acts to the workplace accidents is undeniable worldwide, research aimed at reducing this factor should be considered a discipline of great interest for society in general.

6. Properties of materials used or produced

The physical and chemical properties of raw material use and produce are other essential factors that need attention to properly monitor the health and safety of workers at workplaces. To do this, detailed attention must be given to workplace Material Safety Data Sheet (MSDS). The importance of MSDS in handling worker exposure to hazardous materials in every workplace cannot be substituted for another record keeping data sheet for the purpose of workers' health and safety. According to Translink online website, MSDS is a document that contains information on the potential health effects of exposure to chemicals, or other potentially dangerous substances, and on safe working procedure when handling chemical products. It is an essential starting point for the development of a complete health and safety program. MSDS contains hazard evaluations on the use, storage, handling, disposal information, and emergency procedures related to that material. Therefore, it is imperative to consider the use of MSDS in every workplace to ensure better control of hazardous materials effect on the employees.

This book highlights significant factors contributing to workers' poor health and safety at work, which could result in injury, chronic diseases, and death. Both theories and practical applications are discussed in this book for easy understanding of the readers.

Author details

Bankole K. Fasanya
Environmental Health and Safety Concentration, College of Technology, Purdue University Northwest, USA

*Address all correspondence to: fbankole@pnw.edu

References

[1] Hämäläinen P, Takala J, Kiat TB. Global estimates of occupational accidents and work-related illnesses 2017. WORLD. 2017;**2017**:3-4

[2] AFL-CIO Safety and Health Office. Death on the Job, the Toll of Neglect: A National and State-by-State Profile of Worker Safety and Health in the United States. Available from: https://aflcio.org/sites/default/files/2019-04/DOTJ2019nb.pdf [Retrieved: 18 March 2019]

[3] Garcie-Herrero S. Working conditions, psychological, physical symptoms and occupational accidents. Bayesian network models. Safety Science. 2012;**50**(9):1760-1774

[4] Wickens CD, Gordon SE, Liu Y. An introduction to human factors engineering. In: Charleston SC, editor. Revision 1. 3rd ed. USA: CreatSpace; 1998

[5] Fasanya BK, Shofoluwe M. Occupational ergonomics: Emerging approaches toward improved worker productivity and injury reduction. In: International Conference on Applied Human Factors and Ergonomics. Cham: Springer; 2018. pp. 385-395

[6] CETENA SpA. Ergonomics and Human Factors. Available from: http://www.cetena.it/en/consultancy/ergonomics-human-factor/ [Retrieved: 18 March 2020]

[7] McCauley-Bush P. Ergonomics: Foundational Principles, Applications, and Technologies. Boca Raton, FL: CRC Press; 2011

[8] Zoltán I. Ignaz Philipp Semmelweis, German Hungarian Physician. 1998. Available from: https://www.britannica.com/biography/Ignaz-Semmelweis [Retrieved: 18 March 2020]

[9] Prasad A, Chok HN, Wilkes L. Hand hygiene practices amongst patients. International Journal of Infection Control. 2017;**13**(2):1-8

[10] Burnett E. Perceptions, attitudes, and behavior towards patient hand hygiene. American Journal of Infection Control. 2009;**37**(8):638-642. Available from: https://doi.org/10.1016/j.ajic.2009.04.281

[11] World Health Organization. Evidence of Hand Hygiene to Reduce Transmission and Infections by Multidrug Resistant Organisms in Health-Care Settings. USA: World Health Organization; 2009

[12] Lai M. Hand hygiene- effective way to prevent infections International. Journal of Current Research. 2015;7(03):13448-13449

[13] Sorensen G, Peters S, Nielsen K, Nagler E, Karapanos M, Wallace L, et al. Improving working conditions to promote worker safety, health, and wellbeing for low-wage workers: the Workplace Organizational Health Study. International Journal of Environmental Research and Public Health. 2019;**16**(8):1449

[14] Jonathan GK, Mbogo RW. Maintaining health and safety at workplace: Employee and employer's role in ensuring a safe working environment. Journal of Education and Practice. 2016;7(29):1-7

[15] Holman D, Axtell C. Can job redesign interventions influence a broad range of employee outcomes by changing multiple job characteristics? A quasi-experimental study. Journal of Occupational Health Psychology. 2016;**21**(3):284

[16] Bond FW, Flaxman PE, Bunce D. The influence of psychological flexibility on work redesign: Mediated

moderation of a work reorganization intervention. Journal of Applied Psychology. 2008;**93**(3):645

[17] Beer M, Spector BA, Lawrence PR, Mills DQ, Walton RE. Managing Human Assets. New York: Simon & Schuster, The Free Press; 1984

[18] Armstrong M. A Handbook of Human Resource Management Practice. 10th ed. London: Kogan Page Limited; 2006

[19] Hayes SC, Luoma JB, Bond FW, Masuda A, Lillis J. Acceptance and commitment therapy: Model, processes and outcomes. Behaviour Research and Therapy. 2006;**44**(1):1-25

[20] Bratton J, Gold J. Human Resource Management Theory and Practice. London: Macmillan Press; 1999

[21] Liao P-C, Liu B, Wang Y, Wang X, Ganbat T. Work paradigm as a moderator between cognitive factors and behaviors: A comparison of mechanical and rebar workers. KSCE Journal of Civil Engineering. 2017;**21**(7):2514-2525

[22] Dodoo JE, Al-Samarraie H. Factors leading to unsafe behavior in the twenty first century workplace: A review. Management Review Quarterly. 2019;**69**(4):391-414

[23] Mason S. Procedural violations—Causes, costs and cures. Human Factors in Safety-Critical Systems. 1997;**1**:287-318

Chapter 2

Women, Ergonomics and Repetitiveness

Igor Bello

Abstract

A fair comparison of the conditions in which men and women work is inconsistent, since although they are interacting with the same objects, means and conditions of work, there are differences in the way each gender work, so it condition that naturally the productive systems segregate and thus establish jobs typically "feminized" based on the best adaptation of women to repetitive work low load. From a physical and psychological point of view, female workers have greater exposure to low strength, repetitive motion of upper extremities that causes gender disparity with its health consequences. This chapter documents a study where females were found to have lower biomechanical negative effects in the upper extremities compared to similar male exposures and a higher rate of productivity, especially in tasks of low force demand. This can be attributed to the fact that men used more strength than what was strictly necessary to accomplish the task, mobilizing a greater number of muscle groups than women; females also showed a greater resilience to conditions of high repetitiveness that demanded high-quantitative psychological demands and still maintain productivity rates over time, evidencing also lower rates of rotation and absenteeism caused by musculoskeletal disorders.

Keywords: occupational risks, occupational diseases, cumulative trauma disorders, musculoskeletal system, gender ergonomics

1. Introduction

Biology has always been a determining factor in the definition of the social roles of men and women. From the earliest social organizations of *Homo sapiens*, women oversaw all aspects associated with reproduction: the raising of children, the maintenance of the domestic space, and the care of the elderly. Men were in charge of productive work, study, politics, and laws.

From a physical point of view, differences in gender structure and strength also marked important work differences, giving women roles that involve more repetitive tasks of low strength with upper limbs (basic agricultural tasks, food preparation, clothing, etc.), while the man took on the role related to tasks of handling of loads, walk long routes, and intensive application of force (hunting, fishing, war, etc.) [1].

According to Scott [2], the separation history between home and work emphasizes the functional and biological differences between women and men that end up legitimizing and institutionalizing these differences as the basis of social organization; this division of labor has been normalized.

"Sexual division of labor is universal, but it is specific to each society, and there is a long cultural variability, which shows that the link between work and gender depends on both cultural and biological differences between men and women" [3].

With the industrial revolution and Taylor's theories about the scientific organization of labor, mechanization and the scarcity of postwar male labor were circumstances that encouraged the intensive inclusion of women in manufacturing, beyond the sexual division of labor based on reproductive tasks. Female labor dominated many industries in the textile, electrical parts, and food sectors, in principle because of need, but in a second phase, due to the positive results found in terms of productivity in highly repetitive tasks [4].

This trend continues to be reaffirmed in the twenty-first century where the scenario of the manufacture of low-weight products, or sub-processes of small parts in more complex products, is carried out by the work of women. Especially in the Latin America region can be found many examples of industrial processes with a marked feminine predominance, such as the Mexican/Central American maquila or the textile production cooperatives in Brazil [5].

Then, some questions arise in relation to women, health, and repetitive work:

- For some jobs that involve repetitive tasks, companies prefer to hire women. Is this because of cultural gender considerations or factors such as increased productivity, availability, or showed efficiency?

- At an international level, there is a higher prevalence of MSD in the female population than in the male population. Is it because women are more susceptible, or are they more exposed than men?

- Are women physically and psychologically more suitable for repetitive work with lower marginal physical loads than men?

This chapter aims to provide experimental scientific evidence through the systematic review and application of valid instruments on this subject, from a perspective that considers the conditions of work, the health of women workers, and business productivity.

2. More gender balance, more productivity?

The World Bank also supports the thesis that companies would be more productive with more women. It does so through a report published in 2014 by the International Financial Group (IFC, of which the World Bank is a member). The report, entitled "Investing in Women's Employment: Good for Business, Good for Development" [6], emphasizes that investment in women's employment has led to greater productivity, greater employee loyalty, and a greater access to talent. It also presents concrete examples of how initiatives adapted to women (training, childcare support, health services) can improve business performance.

But this labor reality, partly derived from the cultural division of labor and partly justified by the achievement of positive results in terms of productivity, has consequences in the health of working women. Although it may seem anachronistic, among the bases of the current work organization, we continue to find the Taylorist concepts that have been evolving in the ultra-specialization of workers to perform simple and highly repetitive tasks (conditioning and muscular learning) and introducing concepts such as variable compensation and production bonus [7].

Exposure to repetitive work has been systematically studied in the European Union and especially in Spain [8]. In the EU, 45% of workers declare to perform monotonous tasks, and 37% declare to perform repetitive tasks. In Spain, 64% of workers say they perform repetitive movements during part of the working day. Forty-five percent of the construction workers, 35% of the industrialists, and 30% of those in the services declare to perform them for more than half a day. The most known harmful effects of repetitive work are primarily musculoskeletal disorders of the back and upper limbs. But the repetitive work also has a great relation with another problem of which we cannot disconnect it: the labor stress and its repercussions on the health of the workers.

Repetitive work, besides its ergonomic implications and its more or less direct musculoskeletal consequences, has a central psychosocial significance. In addition to the cyclical realization of the same movements hundreds or thousands of times that forces the maintenance of uncomfortable positions, repetitive work also means a lack of variety tasks, few opportunities for learning, few things to decide, monotony, and boredom [9].

Repeating the same motion can cause injury, but injuries can take years to manifest, as opposed to an accident where the injury is seen immediately. Also, as chronic problems develop over a long period, many elements can interact. It is not in all cases that repetitive elbow movements result in tendon inflammation as one also has no evidence that women workers have a part-time job and have fewer problems than full-time workers. Musculoskeletal problems by repetition frequently are muscle and tendon inflammations such as tendinitis and bursitis. They may also treat deteriorations of cartilage and bones as in some cases of osteoarthritis and different types of spine and nerve compression problems such as carpal tunnel syndrome. These problems are found in both men and women. But several studies consistently show that women present these problems more frequently, and this situation is repeated in studies on the general population as in studies on different occupational groups [10].

If we follow the traditional model, we would say biological differences in size, muscular strength, and aerobic capacity in combination with a very demanding work are sufficient causes to explain these differences. In the last decade, the psychological characteristics of the individual and the psychosocial environment are discussed as causal factors that can modify the response of the organism. These factors may confuse interpretation of research data, turning more difficult to find relationships between causes and effects on health.

3. Why women have more MSD in upper limbs than men?

Zwart [11] and his colleagues collected data from a questionnaire answered by a large sample of German workers, men and women. For their analysis, they divided the data by age and the demand at work (heavy or light physical demands, mental demands, or mixtures of physical and mental demands). Data shows that men and women are affected by back problems, but women refer more frequent problems in the neck and upper limbs. Reviews performed by NIOSH [12] and Artazcoz [13] show several studies with the same situation.

A first approach to explaining these differences in women's working conditions and their exposure to risk factors are based in several beliefs, not all of which are sufficiently well-founded: smaller people have more problems, because tools are too big for small hands; the duration of work (in years of service) with a higher-risk exposure; women have less muscle strength than men, the same conditions have greater effects on them; family responsibilities combined with working conditions increase the risk; hormonal factors alone or in interaction with working conditions produce a greater risk; and women express their problems more psychologically, so women react more to organizational factors that combine with physical factors to produce musculoskeletal problems [14].

However, these explanations are apparently logical and do not support a more in-depth analysis. The most relevant factor is that it can make direct comparisons between the tasks performed by men and those performed by women. Men and women occupy different spaces in the labor market, a way to mark that one could almost speak of separate labor forces. In Sweden, only 10% of women and men work in mixed jobs [15]. This segregation means that men and women are not exposed to the same working conditions.

Within a single economic sector, women occupy different jobs than men; for example, in women's clothing factories in Canada, women work by sewing and ironing clothes, while men are pattern cutters. In the automotive industry, women sew seat linings, while men handle the assembly of vehicles. Even when men and women occupy the same position, the tasks performed are different. In a turkey meat industry, men and women worked in the same production chain; however, men worry about hanging the animal and making large cuts (descading, large cuts, peeling), while women make small cuts (wings, breast, legs) and did the finishing work (cut off excess fat, clots, etc.).

Men and women have different sizes and so, also different reaches. Because of this, they can perform their tasks differently. Karlqvist [16] in his study showed that women work with the computer keyboard in postures more uncomfortable than men because the length of their arms is smaller. However, after many years of female employment in certain productive sectors, furniture, tools, and workplaces have been modified to the dimensions of their workers (anthropometric adaptations), not only obeying gender differences, necessary for implementing productive technologies in other countries (anthropotechnology). In this way, the jobs have been adapted over time to the working population, so that in the positions traditionally occupied by women, the physical configuration is mostly adapted, so these positions could be more injurious to a population with different anthropometric characteristics, including male labor population.

Another reason that may explain the differences between men and women is the work history at long term. Torgén et al. [17] showed that on average, women did not change their physical workload over a 24-year period, while men gradually decreased it. It seems that women stay longer in the same jobs, and it exposes them more time to the same risk factors. Also, when the damage occurs, female workers develop strategies to avoid pain and to be able to work. When men are in these situations, they are more likely to drift toward other positions. The proportion of men doing repetitive work in slaughterhouses decreases with age, while age does not influence the proportion of women in repetitive work. What it probably wants to say is that men withdraw from those jobs.

Other factors that increase the risk of musculoskeletal problems are the lack of rest. Women combine wage labor with housework and complain more often that they are fatigued. Most researches show that women spend more hours on domestic, unpaid work and that this increases with the number of children in the family [18]. The physical and mental demands of domestic/unpaid work are stronger for women than for men (**Figure 1**) and may increase the risk of musculoskeletal problems. This dual presence, both physical and psychosocial, increases the exposure of women to the effects of work and creates an imbalance between the productive and reproductive periods of work and rest that has an impact on the health of women. However, this factor also refers to the greater resilience of women to physical and mental stress caused by work.

Women report more symptoms than men regardless of the region of the body being studied. For example, in a review of computer work and musculoskeletal problems, Punnet and Bergqvist [19] showed that women report more complaints but have fewer

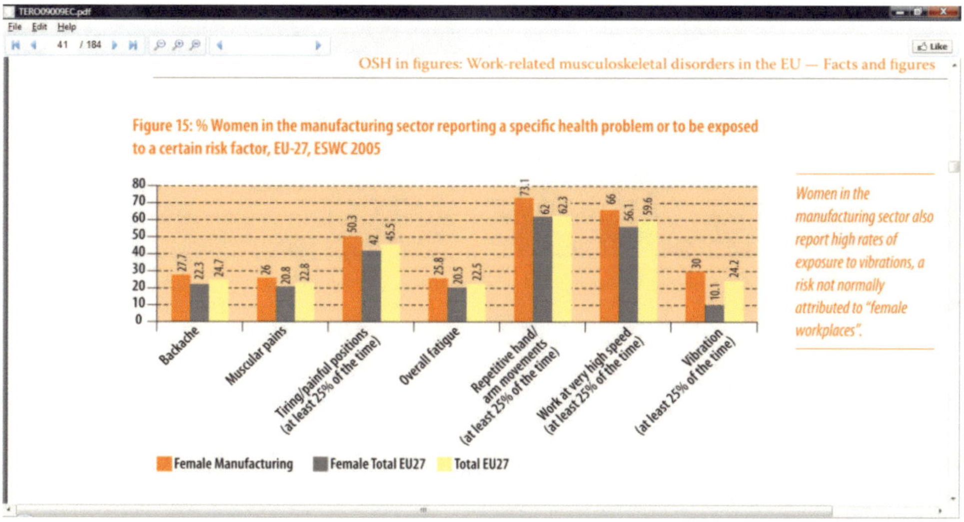

Figure 1.
Percentage of women workers reporting a specific health problem, by risk factor exposition, EU-27, ESWC 2005.

clinical diagnoses than men. One explanation may be that the perception of pain differs between men and women. The perception of pressure pain is more developed in the woman. It could be seen as a weakness, but in reality it proves to be a protective factor which produces a preventive action from the woman at the first symptoms (could be a pre-pathological stage) that induces her to change the way of working.

An experimental, cross-sectional study was conducted between 2015 and 2016 with a semi- probabilistic sample composed of 300 workers (150 women and 150 men) from three industrial plants in Valencia, Venezuela; Santa Cruz de Aragua, Venezuela; and Tijuana, Mexico, in sectors of electrical fan manufacturing, snack packaging, and assembly of hydraulic connections (**Figures 2** and **3**).

The study correlates three variables, considering the biomechanical performance in terms of force and movement range: job hazards (biomechanics, repetition and postures; psychosocial, quantitative psychological requirements), workers' health assessment, and average labor productivity.

Biomechanical risk for repetitiveness was assessed by the OCRA index method [20]. Postural risk was assessed using the OWAS method [21]. Psychosocial risk was assessed using the ISTAS/COPSOQ method [22], specifically through the section on quantitative psychological requirements. The health assessment focused on the

Figure 2.
Gender differences in daily paid/unpaid work. Global Gender GAP Report 2017.

Figure 3.
Snack packaging (Venezuela) and hydraulic connections (Mexico) assessed processes.

functional status of the upper limbs, especially the wrists and forearms (search for initial states of De Quervain pathology, carpal tunnel syndrome, trigger fingers, bursitis, etc.) it performed using dynamometry and goniometry, and also applying Kuorinka questionnaires [23] to identify painful symptoms. We also tested the functional status of the legs, required by the posture of sitting and prolonged standing. Health outcomes were assessed according to the number and severity of injuries. The level of mental load impairment was assessed using the Maslach Burnout Inventory [24]. Productivity was extracted directly from the payroll system (production bonds).

Hand dynamometry (**Figure 4**) assessed the biomechanical performance to measure the interaction between the hand and objects, making a follow-up by day and week. Also, the workflow was video recorded to assist the repetition counter and to apply the measure of movement arches, using a computer-based photo-goniometer.

Variables were correlated, and it got several odds. First, conditions of biomechanical risk were correlated with the occurrence of musculoskeletal disorders. Second, mental load at work, correlated with the emotional exhaustion of the subjects. A third correlation analysis was carried out on the combined effect of biomechanical and psychosocial risks with the occurrence of musculoskeletal disorders, by gender. The synergetic effect of physic and mental work load was assessed through the performance analysis to state cause-effect relations.

Finally, a longitudinal analysis of the productivity (including absences due to rest and rotation of personnel) of each gender was performed, from the perspective of marginal returns and the Cobb-Douglas [25] model curve.

Women were found to have had a lower biomechanical involvement in the upper limbs in the presence of a similar exposure to the male, having, on average, a higher rate of productivity, especially with low force demands, performing repetitive tasks involving a prehensile effort of less than 15 N or manual loads of less than 50 Gr per repetition.

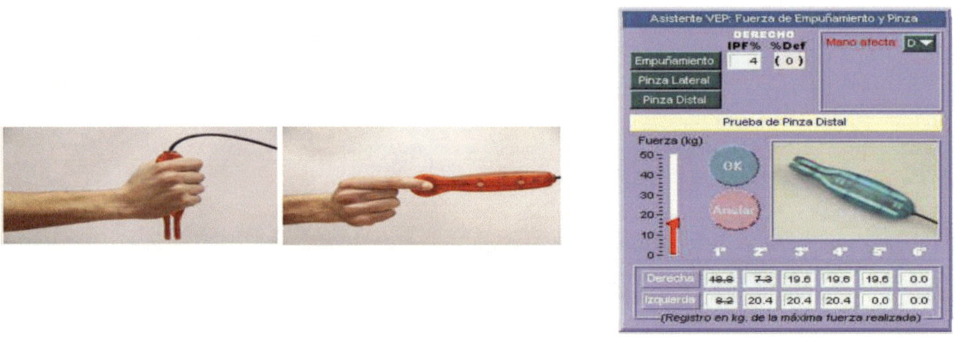

Figure 4.
PC-based hand dynamometer used (probe detail).

It was found that the working conditions (biomechanical and psychosocial) were the same for each gender group in each work facility, but there is a difference that should be considered in the results, since the frequency of technical actions (repetitive movements) was slightly higher in the female group than in the male group, for which the OCRA indexes were also higher in the group of women, and a higher exposure risk was found.

With more repetitiveness, the mental load is also higher, so there is also a greater psychosocial risk in the group of women. The evaluated work stations denoted conditions of high monotony associated with drowsiness, psychic laxity, decreased performance, reduced adaptability, low reactivity, and high variability of heart rate. Hypovigilance conditions were also observed, accompanied by reduced performance in terms of perception and detection of signals, fundamentally in light monitoring tasks. Thus, for both groups, a similar influence of external factors (mental contraite) was found but with a marked difference in the way it affected genders (mental astreinte).

In terms of musculoskeletal affections, greater presence of biomechanical and psychosocial risk would be expected. However, although the pain perception was slightly higher in the female group, the health evaluation found a greater number of lesions in the male group, some with functional limitations.

Among the causes of the observed phenomena, it can be attributed to the fact that men used, on average, more strength than strictly necessary for the accomplishment of the task (which was evidenced with the dynamometer), mobilizing a greater number of muscle groups than those used by the female group.

Another observed phenomenon is that the male group always started the day (weekly and daily) with peaks of productivity (higher than the average productivity of the female group), but this was decreasing throughout the day and week; however, the productivity of the female group remained more constant and finally was, on average, slightly higher than male group.

The group of women also showed a greater resilience to conditions of high repetitiveness that impose high-quantitative psychological demands and still maintain productivity rates over time, which did not happen in the group of men, evidencing also lower rates of turnover and absenteeism due to musculoskeletal disorders.

Women showed a more constant productive rate over time, since the repetition factor hardly changed during the weekly lapses, while the male group showed a marginal productivity distribution similar to the Cobb-Douglas (**Figure 5**) function.

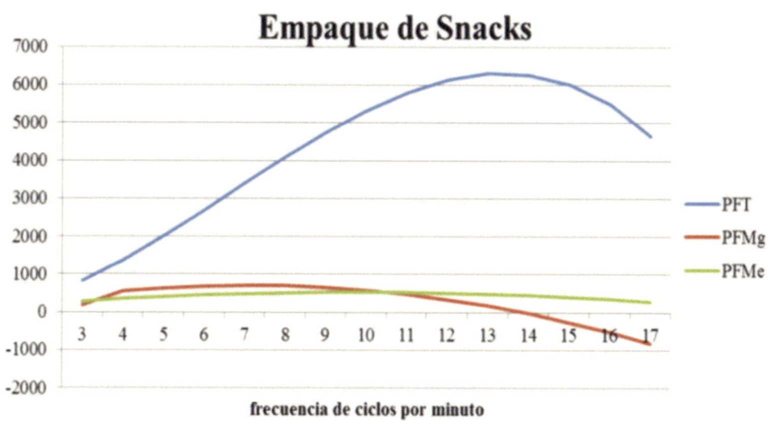

Figure 5.
Cobb-Douglas curves: total product, marginal product, and average product.

A more longitudinal analysis of the marginal productivity per worker in the last year shows even more definitive results, with the female sample having an annualized productivity 18.3% higher than the male sample, considering the effect of absence, rest, and labor rotation.

4. Conclusions and final ideas

The comparison between men and women in labor terms is not fair. Both genders perform different tasks, with different expositions, different patterns of rotation, and different resources to face the work. Probably, both gender are based on a differents patterns but complementaries.

Several studies show a major biomechanical affectation in female group than male group of workers in a great variety of industrial tasks [26]. But at least performing repetitive tasks involving a prehensile effort of less than 15 N or manual loads of less than 50 Gr per repetition, these results are not caused by a gender weakness but a higher-risk exposition in both ways: in short-term time, because women perform more repetitions monthly than men; and in long-term time, because they can support repetitive loads of work for longer time, so they remain for longer time in the repetitive job workplaces.

These results are consistent with Tomaskovic-Devey [27], which results show that exposure to repetitive work is more frequent among working women; and it is consistent with Kilbom [28], whose results show that a greater proportion of women than men suffer problems especially in the upper limbs; finding out that occupations at high risk of musculoskeletal problems in upper limbs, the majority had a mostly female workforce. Results are consistent with NIOSH too, showing that "female gender was a significant predictor of tendinitis."

Although the incidence and prevalence for some authors have not yet been established for De Quervain's disease, some report prevalence between 2.5 and 8% in women of the total working population. Women are more frequently affected than men (8:1 ratio), and the age of onset is between 30 and 60 years [29].

Hobby [30] tested the effects of age and gender on the results of surgery for carpal tunnel release in 97 patients with a diagnosis of CTS. Evaluation was performed before the release and 6 months after surgery. Women reported more symptoms and postoperative disability than men.

These researches show that repetitive work is more frequent among feminized positions, and musculoskeletal disorders in upper limbs are more frequent, pointing to an obvious correlation. But these studies do not inquire about the technical justifications for these phenomena and only describe them. This chapter is not intended to delve into the results of prehensile strength, acceleration of movements, or comparative productivity between genders (which are found in more detail in another article by the authors [25]) but to express conclusions that may be useful to understand probable causes of the phenomena studied and thus explore measures that minimize the effects of repetitive work in women.

This study does not show a greater biomechanical resistance of the feminine articular systems in relation to the masculine ones, but it shows a greater "muscular intelligence" in the accomplishment of the repetitive tasks, which acts as a protective factor and enhances productivity. This feature is based on energy savings and biomechanical resources in each technical action, applying the pressure and acceleration and moving just the required to complete the task. These savings permit performing repetitive tasks for a longer time and reducing the damage to articulations. Performing repetitive actions in a healthy way, without damage, have several productive consequences: taking a maximum advantage of the learning curve, with a better adaptation to muscular/cognitive level and with notable effects in productivity.

There is another implication of the findings, which is related to the validity of sampling in risk assessment. OCRA has proven to be a valid and reliable method for assessing the risk of repetitive tasks; however, one cannot speak of an index of risk associated with the task of, for example, packaging. Each worker has a different rhythm, so the frequency multiplier changes. But it not only changes from worker to worker but also changes the factor at the beginning of the day than at the end of the day and also within each week.

Another disruptive factor is the strength, acceleration, and the way the movements are made. OCRA considers technical actions, but this can be done by mobilizing different muscle groups, with significant variations of strength, acceleration (force is related to mass and acceleration), and associated fatigue. Therefore, repetitive work evaluations, especially when establishing work-gender relations, should be done by gender-based and serial sampling.

One of the most interesting results in this research is related to the different ways men and women face repetitive work. It is clear that women could maintain constant work rhythms, due to better energy management in terms of strength and ranges of movement. However, the group of men showed consistent behavior when starting cycles with greater force but applying less force throughout the day and the week. The range of motion in the male group behaved inversely proportional to that of the applied force, showing ranges of motion and activation of a greater number of muscle groups as the force diminished, in the time. Biomechanically, this situation can be explained as a physical compensation, which uses more muscle groups than necessary to assist fatigued groups. It is not clear why this happens. The answer seems to be not related to the biomechanical sphere, but to a series of conditioning factors of human behavior, determined by psychological, social, and cultural issues. It was not within the scope of this research to deepen the identification of these conditions, but it seems a very promising line of research that can be derived from these findings.

What seems to be clear is that it is not due to special anatomical conditions, although the development of "muscle intelligence" may be linked to a mechanism of evolutionary adaptation, since historically the woman has been performing tasks more repetitive than the man attending to the theory of the cultural division of labor [31]. This adaptation could also intervene in the mental sphere. Thus, from a macroergonomic point of view, the female population was more resilient: able to sustain the work rhythms over time, which allows increasing labor availability and human reliability and minimizing turnover.

Then, a fair comparison of the conditions in which men and women work is inconsistent, since although they are interacting with the same objects, means, and conditions of work, the way each gender work is different, and it conditions that productive systems self-segregates naturally, establishing jobs typically "feminized" based on the best adaptation of women to low-load repetitive work, so this segregation seems to be justified for productive reasons, and not cultural in all cases.

However, if these protective factors are identified, it can be incorporated into the working procedures as part of an action plan to prevent the ergonomic and psychosocial risks from which both genders can benefit.

Acknowledgements

Author wants to thank the collaboration for this chapter to the following institutions:

- Human Engineering Center of Simon Bolivar University (CIH-USB), by the support with lab facilities and data analysis
- Biomechanics Institute of Valencia (IBV), Spain, by all the biomechanics measurement technology used
- Scientific Committee on Women, Health and Work of the International Commission on Occupational Health (ICOH), by the institutional support

Conflict of interest

The author declares no conflict of interest related to funding, commercial intentions, or any other nature.

Author details

Igor Bello
Scientific Committee on Women, Health and Work of International Commission on Occupational Health (ICOH), Valencia, Venezuela

*Address all correspondence to: ibello.medex@gmail.com

References

[1] Seifert A. El trabajo de la mujer y los riesgos de lesiones músculoesqueléticas. I Foro de Salud Laboral: Lesiones músculoesqueléticas. Centre d'étude des interactions biologiques entre la Santé et l'Environnement (CINBIOSE). Universidad de Québec. 1999. Available from: http://www.istas.ccoo.es/descargas/bajar/iforo4.pdf [Accessed: 06 November 2017]

[2] Scott J. La mujer trabajadora en el siglo XIX. Historia de Las Mujeres en Occidente. Historia de las mujeres en Occidente. Vol. 4. Taurus Madrid; 1993. pp. 405-436. ISBN: 84-306-9823-X

[3] DArgemir DC. Trabajo, género, cultura: La construcción de desigualdades entre hombres y mujeres. Icaria; 1995. ISBN: 9788474262544

[4] Rodriguez Carrasco J. Taylorismo. La Revolución mental que llega a Europa. Madrid: UNED; 2015. ISBN: 9788436269260

[5] Delgado Y. Género y Marginalidad—Mujer y Manufactura en América Latina. Cuadernos Americanos—Nueva Época. Vol. 111. Valencia; 2005. pp. 25-32. ISSN: 0185156X

[6] International Financial Corporation: Investing in Women's Employment: Good for Business, Good for Development [Internet]. 2013. Available from: http://www.ifc.org/wps/wcm/connect/topics_ext_content/ifc_external_corporate_site/sustainability-at-ifc/publications/publication_report_investinginwomensemployment [Accessed: 18 November 2017]

[7] Brunel S, Moreno N. Guía para la mejora de las condiciones de salud y trabajo de las mujeres. [Internet]. 2004. Available from: http://www.istas.ccoo.es/descargas/asesora/Guia%20mejora%20condiciones%20%20salud%20y%20trabajo%20Mujeres.pdf [Accessed: 19 November 2017]

[8] European Foundation for the Improvement of Living and Working Conditions: Ten Years of Working Conditions in the European Union [Internet]. Luxembourg: Officials Publications of the European Communities; 2000. Available from: http://www.uni-mannheim.de/edz/pdf/ef/00/ef00128en.pdf [Accessed: 10 December 2017]

[9] Schneider E, Irastorza X. OSH in figures: Work-related musculoskeletal disorders in the EU. Facts and figures. Luxembourg: Publication Office for the EU; 2010. DOI: 10.2802/10952

[10] Canadian Centre for Occupational Health and Safety: Work-related Musculoskeletal Disorders (WMSDs) [Internet]. 2017. Available from: https://www.ccohs.ca/oshanswers/diseases/rmirsi.html [Accessed: 18 November 2017]

[11] Zwart B, Frings M, van Dijk F. Physical workload and the aging worker: A review of the literature. International Archives of Occupational and Environmental Health. 1995;**68**(1):1-12. PubMed ID: 8847107

[12] National Institute for Occupational Safety and Health: Musculoskeletal Disorders and Workplace Factors. A Critical Review of Epidemiologic Evidence for Work-Related Musculoskeletal Disorders [Internet]. 2010. Available from: https://www.cdc.gov/niosh/docs/97-141/pdfs/97-141.pdf [Accessed: 02 October 2017]

[13] Artazcoz L, Cortés I, Moncada S, Brotons M. Diferencias de género en los factores asociados al dolor de espalda crónico en trabajadores. Archivos de Prevención de Riesgos Laborales. 1999;**2**(1):11-18. Barcelona

[14] Kauppinen K, Kandolin I. Gender and Working Conditions in the European Union. Fundación Europea para la Mejora de las Condiciones de Vida y de Trabajo [Internet]. Dublin. 2012. Available from: https://www.eurofound.europa.eu/es/ef/sites/default/files/ef_files/pubdocs/1997/59/es/1/ef9759es.pdf [Accessed: 12 November 2017]

[15] Domínguez Reyes E, Icaza R, Quintero C, López S. Women workers in the maquiladoras and the debate on global labor standards. Feminist Economist. School of Global Studies, Ibero-American Institute. 2010;**16**(4):185-209. ISSN: 1354-5701

[16] Karlqvist LK, Bernmark E, Ekenvall L, Hagberg M, Isaksson A, Rostö T. Computer mouse position as a determinant of posture, muscular load and perceived exertion. Scandinavian Journal of Work, Environment & Health. 1998;**24**(1):62-73

[17] Torgén M, Nygård CH, Kilbom A. Physical work load, physical capacity and strain among elderly female aides in home-care service. European Journal of Applied Physiology and Occupational Physiology. 1995;**71**(5):444-452

[18] Campos-Serna J, Ronda-Pérez E, Artazcoz L, Moen BE, Benavides F. Gender inequalities in occupational health related to the unequal distribution of working and employment conditions: A systematic review. International Journal for Equity in Health. 2013;**12**:57. DOI: 10.1186/1475-9276-12-57

[19] Bergqvist U, Wolgast E, Nilsson B, Voss M. Musculoskeletal disorders among visual display terminal workers: Individual, ergonomic, and work organizational factors. Ergonomics;**2007**:763-776. DOI: 10.1080/00140139508925148

[20] Colombini D, Occhipinti E. Risk Analysis and Management of Repetitive Actions: A Guide for Applying the OCRA System (Occupational Repetitive Actions). 3rd ed. Milan: CRC Press; 2016. ISBN: 1315352583

[21] Karhu O, Kansi P, Kuorinka I. Correcting working postures in industry: A practical method for analysis. Applied Ergonomics. 1977;**8**(4):199-201. DOI: 10.1016/0003-6870(77) 90164-8

[22] Kristensen T, Hannerz H, Høgh A, Borg V. The Copenhagen psychosocial questionnaire—A tool for the assessment and improvement of the psychosocial work environment. Scandinavian Journal of Work, Environment & Health. 2005;**31**(6): 438-449. DOI: 10.5271/sjweh.948

[23] Kuorinka I, Jonsson B, Kilbom A, Vinterberg H, Biering-Sorensen F, Andersonn G. Standardized Nordic questionnaires for the analysis of musculoskeletal symptoms. Applied Ergonomics. 1987;**18**(3):233-237. DOI: 10.1016/0003-6870 (87) 90010-X

[24] Maslach C, Jackson S, Leiter M. Maslach burnout inventory manual. In: Zalaquett CP, Wood RJ, editors. Evaluating Stress: A Book of Resources. The Scarecrow Press; 1985. pp. 191-2018. DOI: 10.1002/job.4030020205

[25] Bello I, Salmen-Navarro A. Ergonomics and increasing marginal returns. In: Advances in Physical Ergonomics and Human Factors. Vol. 489. Cham: Springer; 2016. pp. 503-514. DOI: 10.1007/978-3-319-41694-6_50

[26] Messing K, Dumais L, Courville J, Seifert AM, Boucher M. Evaluation of exposure data from men and women with the same job title. Journal of Occupational and Environmental Medicine. 1994;**36**(8):913-917

[27] Tomaskovic-Devey D, Skaggs S. Sex segregation, labor process organization, and gender earnings

inequality. American Journal of Sociology. 2012;**108**(1):102-128. DOI: 10.1086/344214

[28] Kilbom Å. Nordic Research on the Elderly Workforce. Vol. 1997. Sweden: Nordic Council of Ministers; 1997. ISBN: 9789289300773

[29] Ministry of Social Protection: Evidence-based comprehensive care guide for musculoskeletal disorders (DME) related to repetitive upper limb movements carpal tunnel syndrome, epicondylitis, and De Quervain disease (GIT-DME). Ministry of Social Protection. Bogota; 2006. ISBN: 978-958-98067-4-6

[30] Hobby J, Venkatesh R, Motkur P. The effect of age and gender upon symptoms and surgical outcomes in carpal tunnel syndrome. Journal of Hand Surgery Scotland. 2006;**30**(6):599-604. DOI: 10.1016/j.jhsb.2005.07.005

[31] Delgado A. Género, salud y trabajo. In: Varias miradas, una sola perspectiva en Mujeres en el Mundo: Historia, revoluciones, dictaduras, trabajo, religión y poesía. Valencia: University of Carabobo; 2007. pp. 201-218. ISBN: 978-980-12-4591-9

Chapter 3

Spinal Deformities with Students in Classroom Teaching in Urban and Rural Areas

Elvira Nikšić and Edin Beganović

Abstract

The goal of this research is to determine the differences between spinal deformities of students in urban and rural areas with regard to the initial and final measurements. This research was conducted on a sample of 1105 students. This research program includes students from the first to the fifth grade in the following 11 elementary schools in Sarajevo. Method of measuring the curvature of the spine based on Napoleon Wolanski's criteria from 1975 entitled D-5 spine posture (ESP). The analysis of spinal deformity of students in urban and rural classrooms was done by using the descriptive chi-square statistics. It has been determined that in the initial measurement, a greater number of examinees with no spinal deformity came from the rural schools. The examinees from the urban schools displayed first- and second-degree spinal deformity. In the final measurement, a greater number of examinees without spinal deformity came from rural schools, 28 (2%) to be precise. There was also a greater representation of first-degree spinal deformity in the examinees from urban schools, 49 (8%) to be precise. However, the representation of second-degree spinal deformity was not present in either of the examinees from rural or urban schools.

Keywords: students, spinal deformities, urban-rural area

1. Introduction

Statistical data show that about 4% of people in Sarajevo have spinal deformities to some degree [1]. This is due to the so-called incorrect body posture that has not been corrected on time. Functional disturbances, which have been caused by incorrect posture, are first noticed on the spine and later also on the chest, legs, and feet. The spine is the main contributor to the upright posture and carries all of a person's upper body weight. That is why the spine is often exposed to many deformities, which can seize the entire spine or just some parts of it. The spine can be contorted in three different ways [1]. The types of spinal distortion include kyphosis, scoliosis, and lordosis. One of the key persons in the prevention and suppression of these deformities is the physical and health education teacher, who has to work on correcting postural disorders [1]. He has to apply the appropriate exercises in order to prevent the incorrect body posture. The exercises should contribute to strengthening all the major muscle groups, which are important in preserving the upright posture and preventing worsening of the deformity. Many authors have dealt with the problem of evaluating body posture, as well as spinal deformities and the reliability of these results [2–5]. The analysis of their work included the postural status of the children in the city of Novi Sad. It was a sample of 242 examinees, who were 6 years old. The evaluation of the postural status of the examinees was conducted with the analysis of certain segments of the body based on the method of Napoleon Wolanski [6]. The analysis of the results showed that the boys had good stomach and spine posture, while the posture of other segments was incorrect. The girls had incorrect stomach and spine posture, while the posture of the other segments was good [6]. At the end, the authors concluded that the postural status of the 6-year olds in Novi Sad was disconcerting, and, in the future, it would require taking appropriate measures in order to prevent and realign those postures to their original state [6]. Using the method of Napoleon Wolanski, an evaluation of the body posture of students in the fifth and seventh grade of an elementary school in Sarajevo was conducted. The results have shown that the students in the seventh grade have significantly worse body posture than the students in the fifth grade. It was concluded that the students in the seventh grade had more poor body posture more often located in the areas of the shoulders and neck [7]. This states that 53% of the examinees have incorrect body posture. In conjugation with this statistic, some authors point out that even more than 70% of the school children have some sort of body disorders and certain difficulties that are a consequence of a lack of body movement [8, 9]. The percentages of the presence of postural disorders in young people substantially vary with different authors, and it depends on the specificity of the sample, age, life area, and the applied methodology in detecting the disorders, etc. Winged blades and concave feet are significantly more present in both the boys and the girls from rural areas than those in urban settings. On the other hand, there is a greater presence of a flat foot deformity in both genders from urban areas. Also, we have noticed that the greatest number of disorders refer to the functional shape, which is at the same time the mildest. It is obvious that the presence of incompletely fixated and structural disorders is significantly smaller. It must be noted, however, that it is a significant percentage in the frequency of the more severe forms of body deviations. It is known that children from rural areas are thinner than the city children for several reasons. Not only are they more physically active, but also they are healthier and receive better nutrition. Considering the fact that the winged blades are a lot easily recognized in thinner examinees than those who have larger amounts of subcutaneous fat, this result is not a surprise. As for the status of the foot arch, the obtained results are only a confirmation that the children from urban areas are less mobile and are not prone to be a part of physical activities, especially those forms that are intended to strengthen the foot and lower leg muscles. The fact is that the greatest percentages of the postural disorders refer to the functional form. However, with an adequate program of corrective physical therapy, we can stop the further progression of the physical deviation into a worse stage and largely correct the postural disorder(s) and align the body to its normal position. Obviously, there is a significantly smaller presence of incompletely fixated and structural deformities [10]. A lot of research has been done on proper body posture [6, 8, 11–13]. In comparison to the urban children, the everyday lifestyle of the children in rural schools requires more physical activity outside during the day and includes a healthier diet. In contrast, the urban children often eat unhealthy food from the school kitchen, such as pizza, croissants, pies, various pastries, etc. The students are still not aware enough to choose vegetables and eat less sweet. Foods with artificial flavors are easily within reach, and, once they taste it, everything else tastes bad. Fast food is a modern convenience and can lead to obesity. This has contributed to improper body posture and deformities in an increasing number of children. It can also produce a series of other health difficulties later on in life. Another reason for these disorders is that urban children have difficulties making new friends. They do not go outside but increasingly use the computer and socialize on Facebook. The research conducted, which compared the problems in children from urban and rural areas, has definitely confirmed that the rural children have less postural disorders of any kind [12]. Today, a variety of programs for physical activity are more available to the school children, which can enhance their quality of life. There are many sports clubs, such as collective and individual sports. While urban areas also have natural resources and water sports as well as extreme sports, the research in this area points to the fact that "a great percentage of elementary school children are not included in sports activities, and that is the reason for a greater presence of incorrect body posture."

2. Working method

2.1 Examinee sample

This research includes N = 1105 school age children, of which 528 students were from urban areas (261 girl and 267 boys) and 577 students from rural areas (281 girl and 296 boys). They attended 11 elementary schools in the Sarajevo Canton and the surrounding area, and their average age was M = 82.864. Students who were from first to fifth grade were tested in every school. The breakdown of students included:

- First grade: 221 students (113 boys, 108 girls)
- Second grade: 214 students (109 boys, 105 girls)
- Third grade: 218 students (107 boys, 111 girl)
- Fourth grade: 237 students (121 boys, 116 girls)
- Fifth grade: 215 students (113 boys, 102 girls)

2.2 Variable sample

In order to determine the spinal status, the following variable was used: the method of measuring the curvature of the spine based on Napoleon Wolanski's criteria from 1975, *D-5 spine posture (ESP)—Evaluation of spine posture*. To determine the spinal status, it was necessary for the examinee to turn his back to the examiner with slightly separated but parallel feet in his normal posture. All of the examinees were in their underwear in rooms where the temperature was from 20 to 23°C.

The description of the measuring instrument—evaluation of spinal posture:

0—Physiological curve: normal and within the sagittal and frontal plane.

1—First-degree deviation: kyphosis, scoliosis, and lordosis.

2—Deviation of some kind: a specific deformity or combination of deformities but on a second-degree level

2.3 Data-processing methods

The analysis of spinal deformities of students in school was done by using the descriptive chi-square statistics. Applying the descriptive statistics, the researchers could determine the numerical and percentage frequency of the spinal deformity. These are displayed in tables and graphs.

2.4 Research trial

The research trial lasted one school year. The trial consisted of the following:

- At the beginning of the school year in September, an initial measurement of the body posture was performed based on the criteria of Napoleon Wolanski (1975), *D-5 Spine Posture (ESP)—Evaluation of Spine Posture*, with the help of a physical and health education professor.

- The examinees exercised according to the corrective exercise program used to prevent and correct incorrect body posture and the mentioned postural disorders (spinal deformities) that were made after the initial measuring (**Table 1**). The corrective exercise program was designed to be conducted through various forms of applied activities within the realms of physical and health education with students as a part of classroom instruction.

Every exercise would start by preparing the body, both physiologically and emotionally. The cardiovascular introduction of the load functions that occurred represented the initial physiological element. The addition of an emotional component into this unique program had an enormous significance on the results. Every exercise which was performed gradually increased in intensity from the easiest to the most difficult. To achieve the best result, the teachers focused on concrete exercise demonstration because it was about strictly defined movements. This is why, after the demonstration and the teacher's explanation, the students tried to do a certain task. The short explanations presented the goal of certain exercises and showed how to perform them according to the age of the child. The content of the program was not static, because various corrective exercises were applied in order to correct and prevent specific spinal deformities (**Figures 1–9**).

The exercises need to be changed and adapted to certain situations depending on students' motivation because certain exercises, if they were repeated daily, became monotonous to the students, and thus they would not pay enough attention to perform the moves correctly.

- After the initial measuring, a program was completed during a 6-month period (which finished at week 31). The program was held during the school year from October to April, when the educators were in the classroom.

- The number of training units consisted of physical and health education two times a week, where corrective exercises were performed in order to prevent and correct the spinal deformities.

- The duration of one class is 45 min.

- At the end of the school year in May, final body posture measurements were conducted according to the criteria of Napoleon Wolanski, 1975 [14], *D-5 Spine Posture (ESP)—Evaluation of Spine Posture*, with the help of a physical and health education professor.

- During the 6-month program, there was no testing and measuring. These were conducted before and after the applied program.

- After the initial and final testing when the measurements were taken, an evaluation was made based on the results.

	Beginning position	Exercise description
Remedial shaping excercises for correcting kyphosis	Beginning position: lying on the stomach	**1.** From the beginning position (hands next to the body), lift the shoulders the shoulder muscles while moving the shoulder blades towards the spinal column. At the same time, lift the head from the ground.
		2. From the beginning position, as in exercise 1, lift the hands, place the palms faced down to the ground and the head faced to the ground.
		3. From the beginning position, the fingers are interlocked resting on the back with the palms backwards. Lift the arms and the shoulder muscles while moving the shoulder blades towards the spinal column.
		4. From the beginning position, the arms are faced downwards as are the palms of the hands. Lift the arms (up the side of the body) and shoulders muscles while moving the shoulder blades towards the spinal column. The head remains in line with the spinal column.
		5. From the beginning position, the arms are on the floor, bent at the elbows, so that lower and upper arms make a 90° angle. The elbows are at the height of the shoulder muscles. Lift the head, the pectoral part of the spine and the arms bringing the shoulder blades closer to the spinal column.
		6. From the beginning position, the arms need to be bent at the elbows with the fingers interlocked resting on the nape of the neck. Lift the elbows as high as possible above the ground, as well as the head and the pectoral part of the spine.
		7. From the beginning position, the arms need to be bent at the elbows, but the hands are on the ground next to the shoulders (a resting support for the arms). At the same time lift the shoulder muscles and the arms and yet twist the hands with the palms facing upwards. This head movement is the same as in exercise 4.
		8. From the beginning position, lift the head as high as possible. The face needs to be looking towards the ground. Interlock the fingers, place the hands on the nape of neck, and use them to create strong resistance.
		9. From the beginning position (with the arms stretched out on the ground in front of the head), lift the arms at the same time turning the palms upwards and moving the shoulder blades in towards the spine.
		10. The beginning position is the same as in exercise 9 where the arms are stretched out in front of the head and on the ground, the fingers interlocked, and the palms turned forward. Lift both arms at the same time. The head, turned toward the ground, follows the movement with the face.

	Beginning position		Exercise description
		11.	From the beginning position as in exercise 10, the arms are stretched and raised from the ground. Bend them at the elbows and rest the palms of the hands on the nape of the neck. Then, stretch them out and bring them back down into the beginning position.
		12.	From the beginning position, positions the arms next to the body, turning the palms so that they face the ground. Lifting the arms, with the elbows slightly bent, bring them above the head. The whole movement takes place above the shoulders muscles, as one lifts the pectoral part of the spine and head. Do the same thing in reverse to get back to the beginning position.
		13.	From the beginning position, the arms are lifted above the head. Lifting the pectoral part of the spine and head, bend the arms at the elbows, like in exercise 5, stretch them out and then bring them back into the beginning position.
		14.	From the beginning position, firmly tighten the stomach and buttocks muscles. Interlock the fingers and place the hands on the nape of the neck using them to create resistance as the head is lifted.

	Beginning position		Exercise description
Remedial shaping excercises for correcting kyphosis	Beginning position: while sitting with stretched knees	**1.**	From the beginning position, with the arms placed on the pelvis, stretch the spinal column supporting and producing resistance with the hands. The head remains extended, upright in relation to the spine, but the shoulder muscles need to be pulled downwards.
		2.	In the beginning position, the arms are behind the body and the fingers are interlocked. Stretch the arms pulling them upwards and back as far as possible while turning the palms backward and bringing the shoulder blades in towards the spine.
		3.	In the beginning position, the arms are stretch and turn outwards. While pulling the upper arm upward, bend the elbows.
		4.	In the beginning position, the arms are raised to the level of the shoulders (up the sides of the body) The upper and lower arm form a 90° angle and are in horisontal position. The palms are pointing towards the ground. In that position pull the arms backwards using only the shoulder joints. Be careful that the elbows do not drop lower than the level of the shoulders.
		5.	From the beginning position the arms are bent the arms and the elbows are at shoulder level. Upper and lower arm need to form a 90° angle. While stretching the pectoral part of the spinal column pull the lower arm as far back as possible.

	Beginning position		Exercise description
		6.	In the beginning position the arms are bent at the elbows, the fingers interlocked, and the palms on the nape of the neck. Pull the elbows back as much as possible not separating the palms from the nape of the neck and the chest needs to protrude forward.
		7.	In the beginning position, the arms are being supported by the palms which are resting on the nape of the neck. The elbows are pulled back. Stretch the arms outwards behind the body, palms turned up and forward.
		8.	In the beginning position, the arms are bent, and the fingers are interlocked with the palms on the nape of the neck. Stretch the arms above the head, turning the palms upwards. Do not unfold the fingers and pull the arms as far back as much as possible.
		9.	In the beginning position, the arms are spread with palms turned forward. Pull the arms backwards, bringing the blades in toward the spine. While performing this move, make sure that the arms do not go below the level of the shoulders.
	Beginning position		**Exercise description**
Remedial shaping excercises for correcting kyphosis	Beginning position: Quadrupedal	1.	From the beginning position, supported by the forearms, lift the pectoral part of the spine to the horizontal position while facing the ground. Extend the arms backwards and lift them up, with the palms turned to the ground.
		2.	From the beginning position (like in exercise 1) Lift the pectoral part of the spine until it is in the horizontal position, interlock the fingers on the nape of the neck with the elbows above the level of the shoulder muscles.
		3.	From the beginning position (like in exercise 1), Lift the pectoral part of the spine until it is in the horizontal position. Once the arms are extended up through the elbows, lift them up and to the side. With the arms stretched in this position, turn the palms upwards.
		4.	From the beginning position (like in exercise 1), lifting the pectoral part of the spine to the horizontal position, the completely extended arms (even through the elbows) need to extend forward and up. The exercise is very difficult so it should only be done a few times in the beginning until the examinee builds up his strength.
		5.	The exercises will be even more difficult in this position if they are done with the help of another student. In the beginning position with the hips above the knees, the body should rest on the elbows as the fingers interlocked on the nape of the neck. Lift the pectoral part of the spine to the horizontal position pulling the elbows above the level of the shoulder muscles.

	Beginning position		Exercise description
	Beginning position: Standing	1.	In the beginning position the arms are behind the body, the fingers are interlocked, and the palms turned backwards. Bending the elbows, pull the upper part of the hands across the back as high as possible. The abdominal muscles need to be kept firmly tightened during the whole movement.
		2.	In the beginning position, the arms are behind the body, the fingers are intertwined, and the palms turned backwards. Pull the stretched arms back as much as possible and up moving the shoulders blades in towards the spine. Pull the stomach in.
		3.	In the beginning position the arms are next to the body. While stretching the spinal column, extend the arms. Perform this movement behind the body as much as possible. With the arms stretched out, turn the palms so that they are going up. In the same way, bring the arms back to the beginning position.
		4.	From the beginning position, raise the arms to a stretched position, bending them at the elbows. At the level of the shoulder muscle, pull the elbow backwards with a pronounced withdrawal of the forearm. Firmly tighten the abdominal muscles and keep the spinal column vertical.
		5.	From the beginning position, lift the arms above the head. With the arms stretched out above, firmly tighten abdominal muscles, draw the stretched arms back as much as possible.
		6.	In the beginning position, bend the upper part of the body to the horisontal position with the face faced down. Pull the arms behind the back with palms facing towards the ground.
		7.	In the beginning position, the hands are on the nape of the neck. Then, bend the upper part of the body up into the horisontal position. At the same time, pull the elbows back as much as possible.
	Beginning position		**Exercise description**
Remedial shaping excercises for correcting lordosis	Beginning position: lying on the back with legs bent at the knees	1.	From the beginning position, pull one knee and then the other one up as close to the chest as possible (alternating them in sequence). Later, in the exercise, do this with both legs at the same time.
		2.	From the beginning position, lift up the head with the face towards the chest without the support of the hands.
		3.	From beginning position, lift and extend both arms toward the knees, lifting both the head and the shoulder muscles. The feet need to remain on the ground.

Beginning position		Exercise description
	4.	From the beginning position, pull the feet up off of the ground. Then, with the knees together, move them from one side to the other side toward the floor.
	5.	From the beginning position, stretch the left leg upwards, pulling it as close as possible to the body core. At the same time, lift only the right shoulder and the head and with your right hand reach for the foot of the extended leg. Then do this process again with the left leg and right hand.
	6.	From beginning position, the hands are interlocked on the nape of the neck. Lift the head and then the shoulder muscles, pulling up the shoulder blades from the ground.
	7.	From the beginning position, bend the legs at the knees and stretch the lower legs.
	8.	From the beginning position, pull the tights towards the chest, then lift the pelvis, so that the knees are above the chin. The arms need to be in an extended position next to the body.
	9.	In the beginning position fingers are interlocked on the nape of the neck, with elbows in the back. Lift the head and the shoulder muscles and rotate them from one side to the other.
Beginning position sitting with stretched legs	**1.**	From the beginning position lift the legs one by one with the knee extended.
	2.	From the beginning position lift both legs at the same time with the knee extended.
	3.	From the beginning the extended legs are being crossed the left leg over the right and the right leg over the left legs (alternating the legs).
	4.	From the beginning position separate the extended legs, pull them back together, and then bring them back to the beginning position.
	5.	From the beginning position lift the extended legs. At the same time pull chest forward and touch the feet with your hands.
	6.	From the beginning position bring the extended upper body backwards. Be careful not to create an increase the curvature in the lumbar.
Starting position: lying on the back with the legs stretched	**1.**	In the starting position, tighten the abdominal muscles and lower the lumbar part of the spine to the ground. In the beginning allow for some knee bending, but later have the examinee do the exercise to their maximum leg extension.
	2.	From the starting position, lift the right arm, right shoulder and head and touch the left hand that is on the ground on the left side of the body. Alternate the arms with this same action having them come cross over the body.

	Beginning position		Exercise description
		3.	From the starting position, (to bend the legs at the knees), pull the up feet along the ground until they're fully bent at the knees. At the same time lift the head until you touch your chest with your chin.
		4.	From the beginning position, lift both legs off of the ground at the same time, crossing them, the left across the right and vice versa.
		5.	In the starting position, the hands are interlocked on the nape of the neck. Lift the upper part of the body to a 45 ° angle.
		6.	From the starting position with the hands interlocked on the nape of the neck, lift the head and the shoulder muscles off of the ground.
Remedial shaping excercises for correcting scoliosis (Right thoracic scoliosis)	Starting position lying on the stomach	**1.**	In the starting position place the arms on the pelvis. While facing the ground, stretch the spine with some intensity, bring the shoulder muscles down.
		2.	In the starting position the left arm is above the head, and the right arm is next to the body. Lift the head and the shoulder muscles together along with the arms along with stretching spine with as much intensity as possible.
		3.	From the starting position the left arm is stretched above the head, the right arm bent at the elbow next to the body. Lift the head and left arm and apply pressure with the forearm on the ground. Later, it can be combined with a deflection to the right side, that should be prevented with the pressure of the right forearm.
		4.	From the starting position do the same as in the exercise 1, with a deflection in the pectoral part of the spine.
		5.	From the starting position the left arm is extended, the right arm is bent at the elbow and that hand rest on the cheek. Lifting the head from the ground, move it toward the right side, which should completely prevent it because of the pressure from the right hand. At the same time, while slightly bending the left arm at the elbow, pull down the hand.
		6.	From the starting position, the left arm is next to the body, with the palm facing down. Another student gives an appropriate amount of resistance on the forearm while the other student tries to lift and bring his stretched arm back.
	Quadrupedal starting position	**1.**	From the starting position, lower the thighs on to the lower legs and have the left extended. The right arm is resting on the forearm on the ground and the elbow a distance from the body. Lift the pectoral part of the spine with left arm extended. The right arm stays on the ground. Repeat this exercise alternating positions of the arms.

	Beginning position		Exercise description
		2.	From the starting position like in exercise 1. Lift the pectoral part of the spine, with arms as they are in the starting position.
		3.	In the starting position lean on the forehead, interlock the fingers and place them on the nape of the neck. Stretching the pectoral part of the spine into the horizontal position, use the hands to produce resistance.
		4.	From starting position, pull the hips back, interlock the hands on the nape of the neck. Produce a deflection in the pectoral part of the spine to the right while at the same time create resistance with the right arm to the movement.
		5.	Sitting with the knees extended could also be used with functional scoliosis localized in the pectoral part of the spine. With a couple of examples, it is possible to choose arm stances given in the previous positions.
Remedial shaping excercises for correcting scoliosis (Left thoracic scoliosis)	Starting position: lying on the stomach	1.	From a starting position make a deflection with the whole upper part of the body to the left. With the left hand touch the left knee.
		2.	In the starting position pull the upper part of the body and the extended leg to the left. The deflection should be in the area of the belt (or waistline). The right leg stays on the ground.
		3.	In the starting position bend both legs at the knees. The thighs and lower legs need to create a 90° angle. Lift both thighs from the ground, not changing the angular relation of thighs and lower legs.
		4.	In the starting position lift both extended legs from the ground. If the scoliosis is not corrected, separate the left leg slightly.
		5.	In the starting position lift the head, and the shoulder muscles from the ground, while the face is facing the ground. Lift the left leg bent at the knee from the ground, and then stretch it out. Lowering the upper part of the body, lower the stretched leg to the starting position.
		6.	From the starting position lift both legs slightly from the ground and pull them to the left.
	Quadrupedal starting position	1.	From the starting position pull the left leg bent towards the chest. The foot should not touch the ground.
		2.	From the beginning position lift the hips until the knee is completely extended and with the ground as a support for the feet. From that position, lift the left leg to the side.
		3.	In the starting position the left knee is in front of the right knee. Extend the right leg completely (including the knee) and lift it to the horisontal position.

Table 1.
Exercise program for preventing and correcting the postural disorders of the spine.

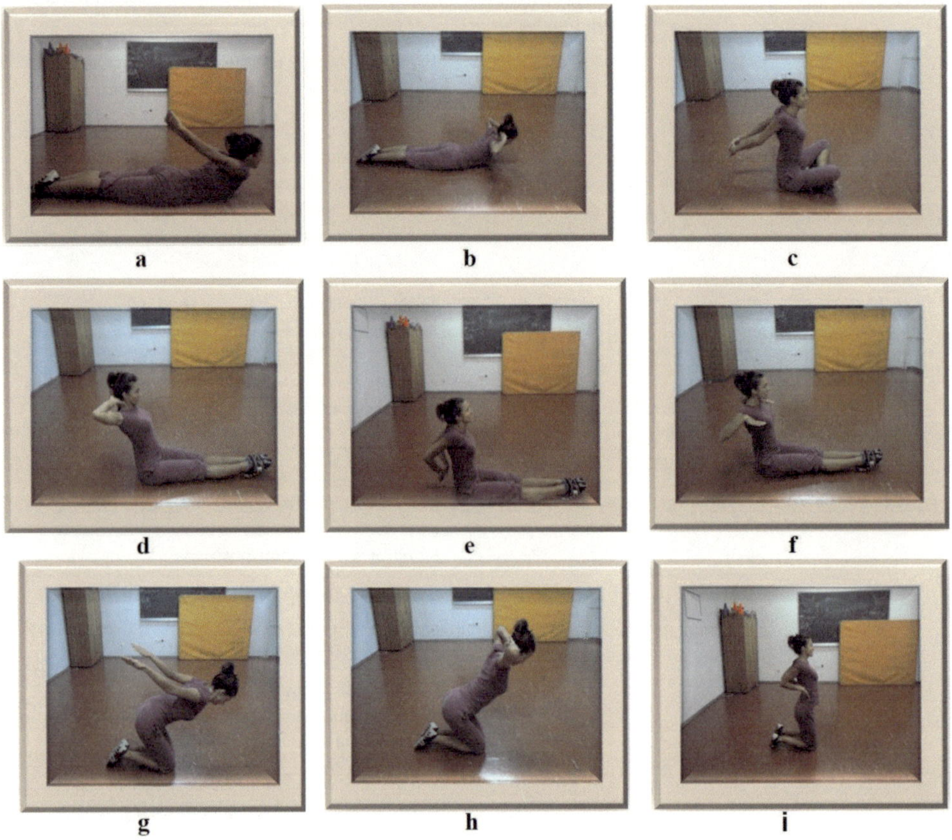

Figure 1.
(a–i) Some of the exercises for the removal of kyphosis.

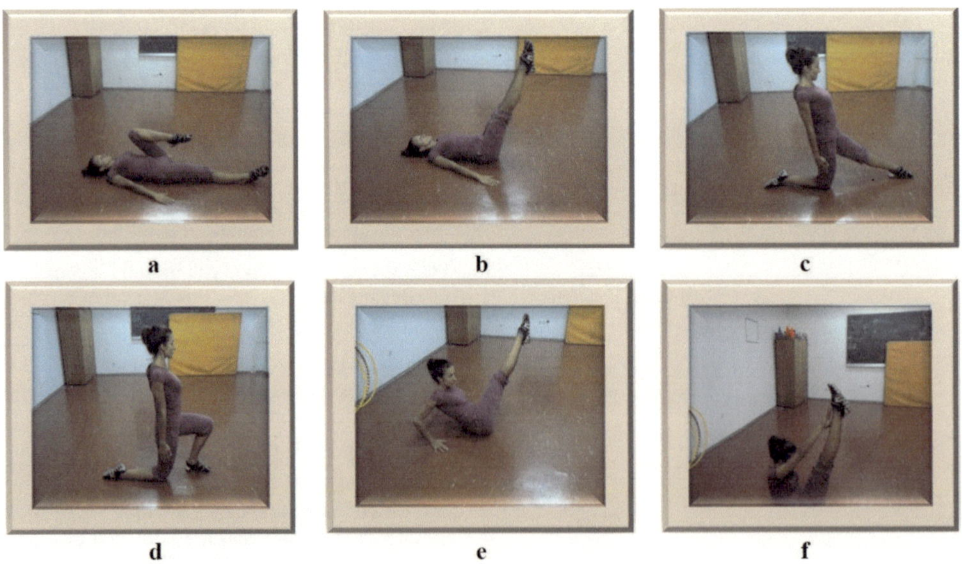

Figure 2.
(a–f) Some of the exercises for the removal of lordosis.

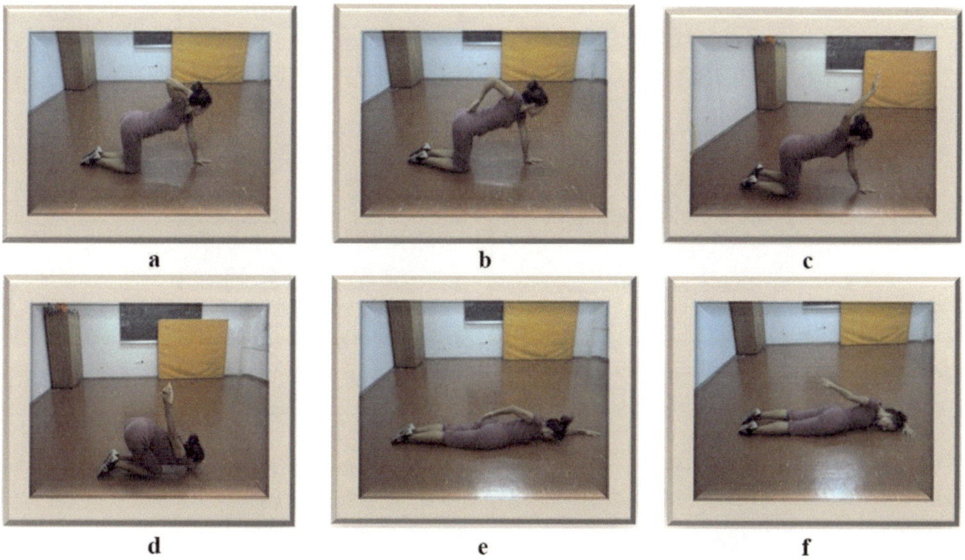

Figure 3.
(a–f) Some of the exercises for the removal of scoliosis.

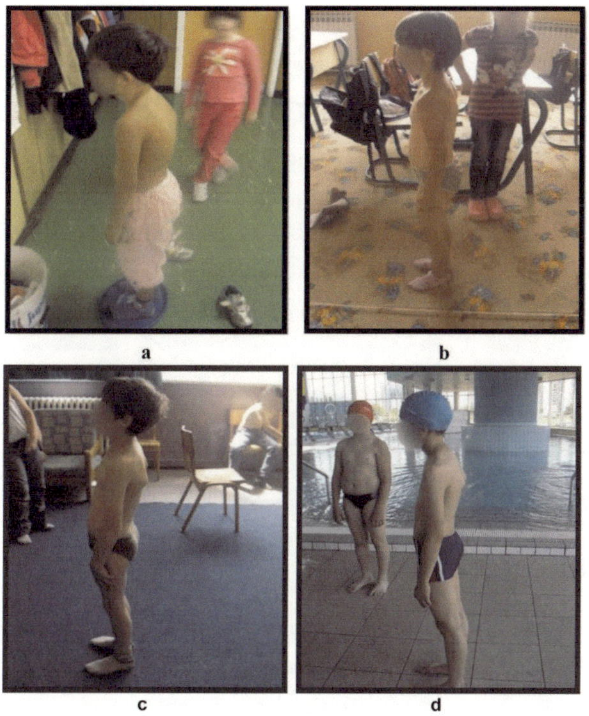

Figure 4.
(a, b) Kyphosis with girls. (c, d) Kyphosis with boys.

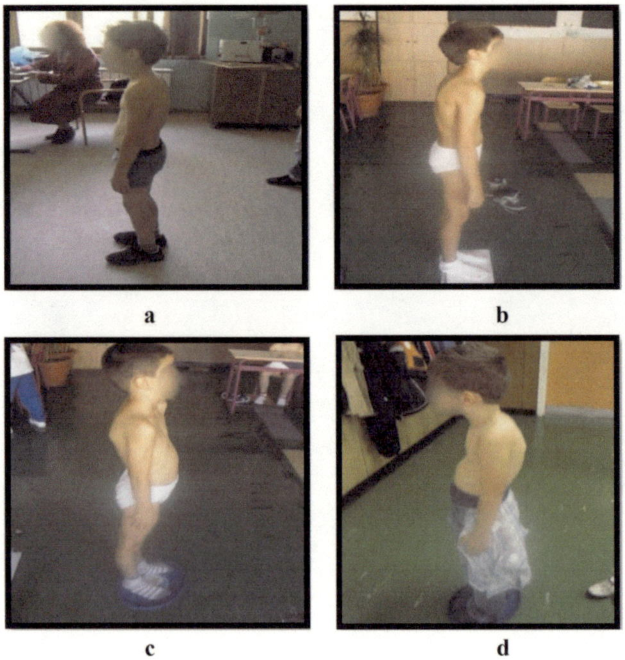

Figure 5.
(a–d) Kyphosis with boys.

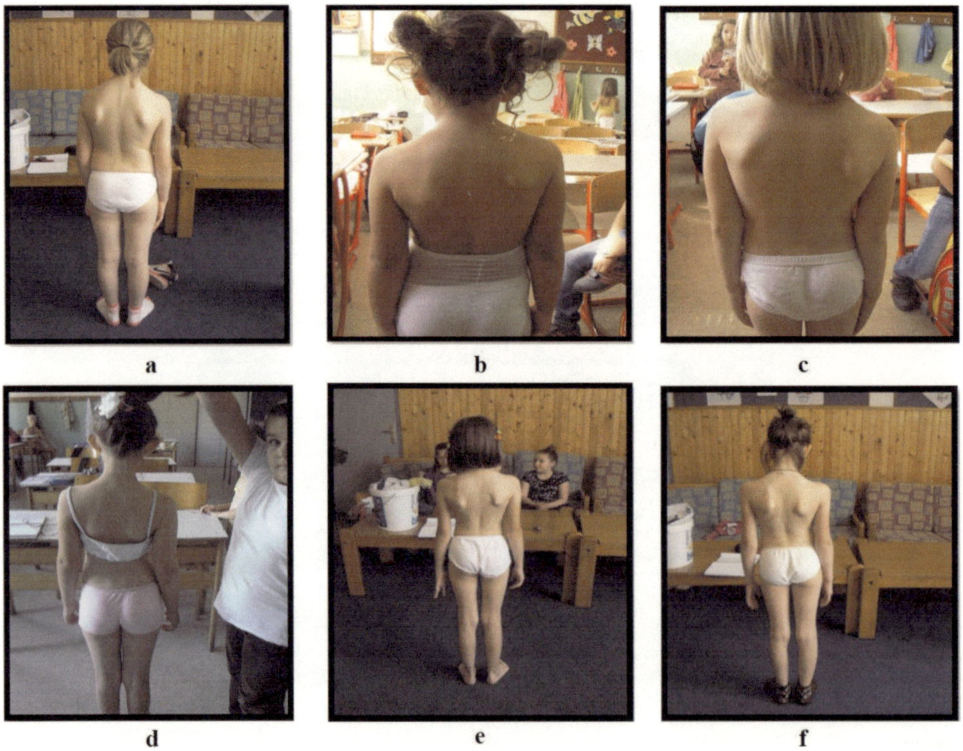

Figure 6.
(a–f) Scoliosis with girls.

DOI: http://dx.doi.org/10.5772/intechopen.87977

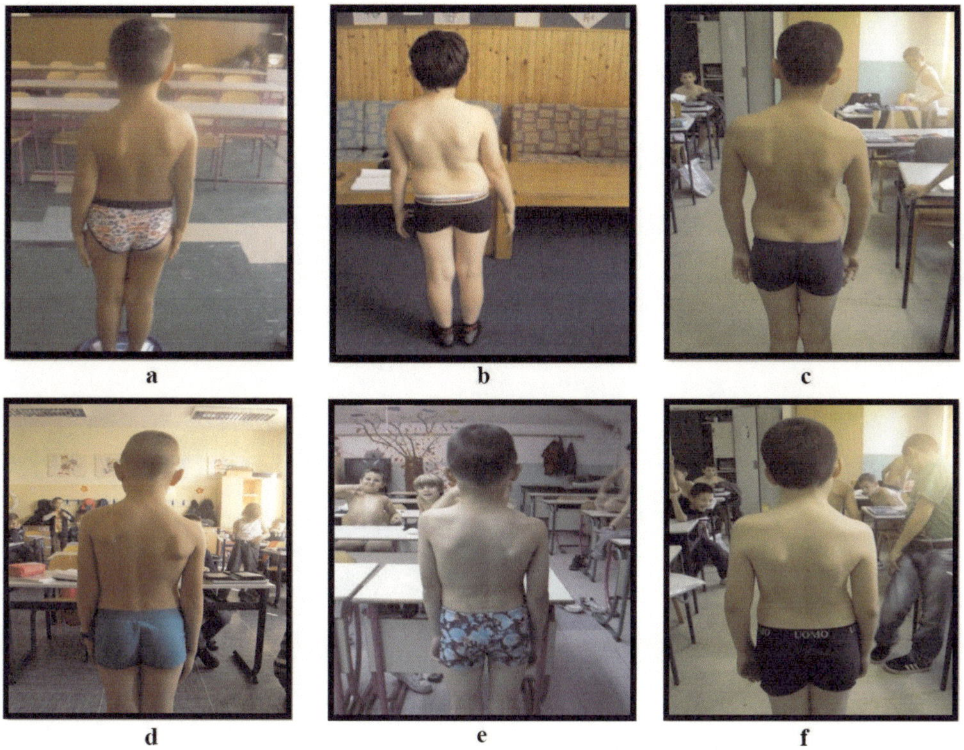

Figure 7.
(a–f) Scoliosis with boys.

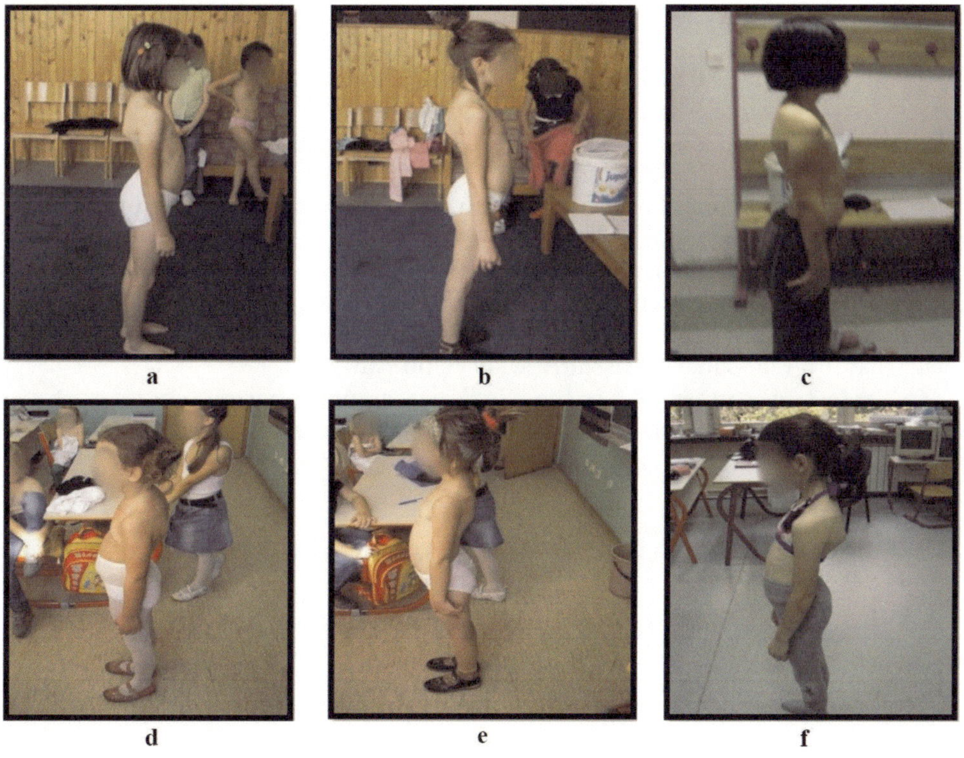

Figure 8.
(a–f) Lordosis with girls.

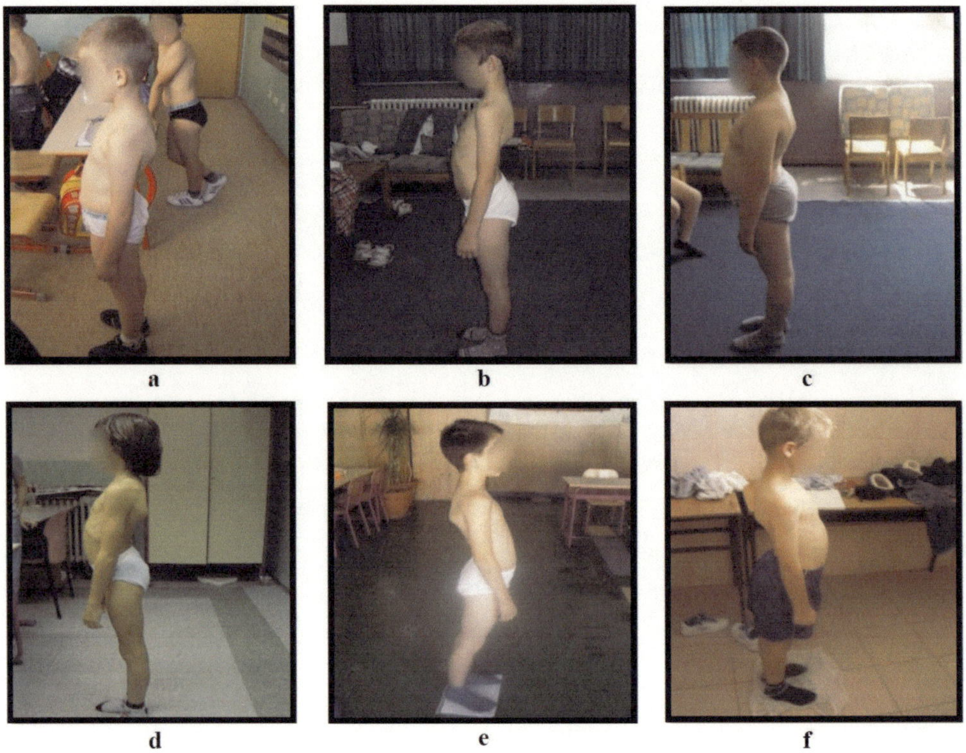

Figure 9.
(a–f) Lordosis with boys.

3. Results and analysis

In this chapter, the data obtained about the frequency and size of spinal deformities of students in classroom are presented and analyzed. The evaluation of spine posture (ESP) is as follows: 0, physiological curve (normal and within the sagittal and frontal plane); 1, first-degree deviation (kyphosis, scoliosis, and lordosis); and 2, deviation of some kind (a specific deformity or combination of deformities but on a second-degree level).

In **Table 2**, we can see that in the initial and the final measurements, there are statistically significant differences between the two genders in urban and rural school settings in the presence and degree of spinal deformities which was on a level lower than 1% (chi-squares are statistically significant).

In **Table 3**, we can see that, in the initial measurement, a greater number of examinees from rural schools had no spinal deformity in comparison to the examinees from urban schools. The examinees from urban schools had a bigger

Body part	Initially			Finally		
	Chi-square	Liberty degree	Level of significance	Chi-square	Liberty degree	Level of significance
Spine	35.750	2	0.000**	54.104	1	0.000**

***Chi-square is statistically significant at a level lower than 1%.*

Table 2.
The values and levels of chi-square significance in examining the differences between urban and rural schools in the degree of spinal posture in the initial and final measurements.

Body part	School	Body posture measuring				
		Grade (number of points)	Initial		Final	
			F	%	F	%
Spine	Rural	0	296	51.3	414	71.8
		1	190	32.9	163	28.2
		2	91	15.8	0	0.00
		Total	577	100.0	577	100.0
	Urban	0	177	33.5	265	50.2
		1	232	43.9	263	49.8
		2	119	22.5	0	0.00
		Total	528	100.0	528	100.0

Table 3.
The distribution of certain grades of spinal posture at initial and final measurements, according the location of the school.

percentage of spinal deformities of the first and second degree. This difference is statistically significant and measures at a level lower than 1%, which is shown by the chi-square from the previous analysis (chi-square = 35.750 and p = 0.000). In the final measurements from **Table 3**, there is a greater number of examinees from rural schools without a spinal deformity, and there is a greater presence of spinal deformity of the first degree in the examinees from urban schools. However, the presence of second-degree spinal deformity is not present in either examinees from rural or from urban schools, and the difference is statistically significant at a level lower than 1% (chi-square = 54.104 and p = 0.000).

In **Table 4** based on the F-ratio values which are significant to the factors of the exercise program, we can conclude that, in the final measurement, there were statistically significant changes in the values of the cervical and lumbar curves. On the other side, both in the initial and in the final measurements, a significant difference was noted in the range of the cervical and lumbar curves given the fact that the difference in size is slight in the final comparison to the initial measuring,

Source of variability	Measure	Sum of squares	Liberty degree	Variance	F-ratio	p
Measurement	Cervical curve	61.595	1	61.595	433,492	0.000 **
	Lumbar curve	13.964	1	13.964	48,869	0.000 **
School	Cervical curve	5.972	1	5.972	9342	0.002 **
	Lumbar curve	19.094	1	19.094	8146	0.004 **
School interaction × Measuring	Cervical curve	1.820	1	1.820	12,806	0.000 **
	Lumbar curve	1.264	1	1.264	4422	0.036 *
Degree of Error	Cervical curve	705.135	1103	0.639		
	Lumbar curve	2585.443	1103	2.344		

**F-ratio is statistically significant on a level lower than 5%.*
***F-ratio is statistically significant on a level lower than 1%.*

Table 4.
The analysis results of the variation in examining the differences between the students in urban and rural schools in the initial and final measurements of cervical and lumbar curves.

which is indicative of the significant interaction from the school and the exercise program.

In **Table 5**, we can see noticeable differences according to the location of the school in the value of the cervical curve in both the initial and the final measurements. Those differences are statistically significant, which is indicated by the value and significance of the obtained F-range (F = 9342 and p = 0.002). Also, there are noticeable differences between the measurements where the values of the cervical curve are smaller than in the final measurement in comparison to the initial measuring. These differences are statistically significant, and they are indicated by the value of F-range from the previous analysis (F = 433,492 and p = 0.000). Additionally, the level of interaction from specific schools involved in the exercise program was determined. So, in the final measurements of the cervical curve, differences based on school location are significantly greater than the initial measuring.

In **Table 6**, we can see that there are differences between the measurements as well as differences regarding the school area in the values of the lumbar curve. The change in the value of the lumbar curve at the final measuring in comparison to the initial measuring is significant for both the examinees from rural and urban schools. This is indicated by a significant F-range (F = 48,869 and p = 0.000). Also, there are noticeable differences regarding the school's location in the initial and final measurements, which is indicated by the significant F-range that takes into consideration the school factor (F = 8146 and p = 0.004). On average, the examinees from rural schools have smaller values for the lumbar curve both in the initial and final measuring. Additionally, the significance of the school's interaction through the exercise program was determined. So, in the final measurements, differences according to gender with regard to the measures of lumbar curve are significantly larger than the initial measurements.

In **Table 7**, we can see that in the initial measurements, there are statistically significant differences between the examinees from urban and rural schools with regard to the presence and degree of the spinal deformity (scoliosis and lordosis).

School	Measuring	N	Smallest value	Biggest value	Arithmetic mean	Standard deviation	Median
Rural	IM	577	1.50	5.00	32.792	0.65634	3.5
	FM	577	1.50	4.10	30.024	0.44999	3.0
Urban	IM	528	1.00	6.50	34.407	0.81478	3.5
	FM	528	1.50	5.20	30.491	0.52858	3.0

Table 5.
The average values and measurements of the deviations in regard to the measurements of the cervical curve at the initial and final measuring, in relation to the school location.

School	Measuring	N	Smallest value	Biggest value	Arithmetic mean	Standard deviation	Median
Rural	IM	577	0.50	8.50	36.312	1.30201	3.5
	FM	577	0.60	5.70	35.199	0.91960	3.7
Urban	IM	528	0.60	8.00	38.652	1.32449	3.5
	FM	528	0.90	6.00	36.581	0.98621	3.7

Table 6.
The average values and measurements of the deviation regarding the measures of the lumbar curve in initial and final measuring, regarding the school area.

Deformity	Initial			Final		
	Chi-square	Liberty degree	Level of significance	Chi-square	Liberty degree	Level of significance
Kyphosis	10.649	2	0.005	16.519	1	0.000**
Scoliosis	18.176	2	0.000**	27.332	1	0.000**
Lordosis	23.042	2	0.000**	16.614	1	0.000**

***Chi-square is statistically significant on a level lower than 1%.*

Table 7.
The values and levels of the chi-square significance in examining the differences between urban and rural schools in regard to the presence of certain kinds of spinal deformities in initial and final measuring.

The chi-square is statistically significant at a level lower than 1%, but there are no statistically significant differences in the presence and degree of spinal deformity (kyphosis). In the final measurement, significant differences are seen between the examinees from urban and rural schools in the degree and presence of the spinal deformity (kyphosis, scoliosis, and lordosis). The chi-square is statistically significant at a lever lower than 1%.

In **Table 8**, it is visible that in the initial measurement of the examinees from rural schools and from urban schools, there are fewer cases of kyphosis, and first- and second-degree kyphosis among those from rural schools. Also, in the final measuring, as seen in **Table 8**, there are a larger number of examinees from rural schools without kyphosis and a larger number of examinees from urban schools with first-degree kyphosis. In addition, second-degree kyphosis was not present in the examinees either from rural or from urban schools.

In **Table 9**, we can see that there are more cases without scoliosis in the initial measurements of examinees from rural schools than the examinees from urban schools. There are also fewer cases of first- and second-degree scoliosis. Also, in the final measurements, as seen in **Table 9**, there are a large number of examinees from rural schools without scoliosis and examinees from urban schools without first-degree scoliosis, while second-degree scoliosis was not present in the examinees either from rural or urban schools.

Spine deformity	School	Body posture measuring				
		Deformity	Initial		Final	
			F	%	F	%
Kyphosis	Rural	Not present	504	87.3	538	93.2
		I degree	56	9.7	39	6.8
		II degree	17	2.9	0	0.00
		Total	577	100.0	577	100.0
	Urban	Not present	425	80.5	453	85.8
		I degree	72	13.6	75	14.2
		II degree	31	5.9	0	0.00
		Total	528	100.0	528	100.0

Table 8.
Display of the frequency distribution of the presence of spine deformity in the initial and final measurements, according to the school location—kyphosis.

Spine deformity	School	Body posture measuring				
		Deformity	Initial		Final	
			F	%	F	%
Scoliosis	Rural	Not present	480	83.2	524	90.8
		I degree	94	16.3	53	9.2
		II degree	3	0.5	0	0.00
		Total	577	100.0	577	100.0
	Urban	Not present	397	75.2	421	79.7
		I degree	113	21.4	107	20.3
		II degree	18	3.4	0	0.00
		Total	528	100.0	528	100.0

Table 9.
Display of the frequency distribution of spinal deformity in the initial and final measuring, according to the school location—scoliosis.

Spine deformity	School	Body posture measuring				
		Deformity	Initial		Final	
			F	%	F	%
Lordosis	Rural	Not present	412	71.4	492	85.3
		I degree	90	15.6	85	14.7
		II degree	75	13.0	0	0.00
		Total	577	100.0	577	100.0
	Urban	Not present	305	57.8	399	75.6
		I degree	130	24.6	129	24.4
		II degree	93	17.6	0	0.00
		Total	528	100.0	528	100.0

Table 10.
Display of the frequency distribution of spinal deformities in the initial and final measuring, according to school location—lordosis.

In **Table 10**, we can see that in the initial measurements, fewer examinees from rural schools have lordosis than the examinees from urban schools, and there are fewer cases of first- and second-degree lordosis. In the final measuring, seen in **Table 10**, there are a greater number of examinees from rural schools without lordosis and fewer examinees from urban schools with first-degree lordosis, while second-degree lordosis was not present in the students either from rural or urban schools.

3.1 Discussion

3.1.1 Table 3

As we were evaluating the spinal posture of the students from rural schools, we came across some data that shows that, in the initial measuring of 296 (5.3%) students, the physiological curvature was normal in both the sagittal and frontal planes, which indicates correct body posture. In 190 (32.9%) students, the deviation of a first-degree deformity was present as kyphosis, scoliosis, or lordosis. In 91 (15.8%) students, there was a singular deviation or a combination of them but on a second-degree level. In 177 (33.5%) students from urban schools, the physiological curvature was normal and within both the sagittal and frontal planes, which indicates correct body posture, while in 232 (43.9%) students, there was a first-degree deviation of kyphosis, scoliosis, or lordosis. In 119 (22.5%) students, there was a singular deviation or combination but on a second-degree level. In the final measuring of 414 (71.8%) students from rural schools, the physiological curvature was normal in both the sagittal and frontal planes, which indicated correct body posture. In 163 (28.2%) students, there was a first-degree deviation of kyphosis, scoliosis, or lordosis, while an expressed deviation, a singular deviation or combination on a second-degree level, was not present in the final measuring of students from rural schools. In 265 (50.2%) students from urban schools, the physiological curvature was normal in both the sagittal and frontal planes, which indicates correct body posture, while in 263 (49.8%) of the students, there was a first-degree deviation of kyphosis, scoliosis, or lordosis, while an expressed deviation, a singular deviation or combination on a second-degree level, was not present in the final measuring in the students from urban schools. Because of this and the tiredness that occurs after sitting too long, children most often take incorrect positions that later grow into bad habits, and in time these become physical disorders of a milder or more severe form. In addition, a child's means to get back and forth to school has changed a lot over the years. This is due mostly to the increase in number of cars. Unlike the previous generations of students who went mostly on foot to school, to practice, to their relatives, etc., children in the twenty-first century are not walking enough. Driving from one place to another exclusively takes place in cars, busses, van, etc. This kind of lifestyle, along with many other conveniences, has certain consequences that have negatively affected the motor capabilities and posture as well as the esthetic look and the quality of life of young people.

3.1.2 Table 8

After measuring the spine curvature of the students from rural schools, the data obtained showed that out of 1105 examinees examined in initial measuring, 504 (87.3%) students had symmetrical shoulders and shoulder blades and a straight spine, which means that they had good body posture. Fifty-six (9.7%) students had the following characteristics: a head bent in the forward position; shoulders bent in the forward position; a pronounced stoop; a recessed chest; protruding shoulder blades; a bulging, flabby stomach; and knees which were slightly bent and shifted forward. This indicates a first-degree kyphosis deformity. Seventeen (2.9%) students with more pronounced symptoms were classified to have a second-degree kyphosis deformity. Four hundred and twenty-five (80.5%) students from urban schools had symmetrical shoulders and shoulder blades and a straight spine, which means they had good body posture. Seventy-two (13.6%) students had the following characteristics: a head bent in the forward position; shoulders bent in the forward position; a pronounced stoop; a recessed chest; protruding shoulder blades; a bulging, flabby stomach; and knees which were slightly bent and shifted forward. This indicates a first-degree kyphosis deformity. Thirty (5.9%) students with more pronounced symptoms were classified to have second-degree kyphosis. In the final measuring, 538 (93.2%) students from rural schools had symmetrical shoulders and shoulder blades and a straight spine, which means they had good body posture. Thirty-nine (6.8%) students had the following characteristics: a head bent in the forward position; shoulders bent in the forward position; a pronounced stoop; a recessed chest; protruding shoulder blades; a bulging, flabby stomach; and knees which were slightly bent and shifted forward. This indicates a first-degree kyphosis deformity. However, the extreme deviation, first-degree kyphosis, was not present in the students from the rural schools in the final measuring. Four hundred and fifty-three (85.8%) students from rural schools had symmetrical shoulders and shoulder blades and a straight spine, which means they had good body posture. Seventy-five (14.2%) students had the following characteristics: a head bent in the forward position; shoulders bent in the forward position; a pronounced stoop; a recessed chest; protruding shoulder blades; a bulging, flabby stomach; and knees which were slightly bent and shifted forward. This indicates a first-degree kyphosis deformity. However, the extreme deviation, first-degree kyphosis, was not present in the students from urban schools in the final measuring.

3.1.3 Table 9

After measuring the spinal curvature of the students from rural schools, the data obtained showed that, out of 1105 examined students in the initial measuring, 480 (83.2%) students had symmetrical shoulders and shoulder blades and a straight spine, which means they had good body posture. Ninety-four (16.3%) students had asymmetrical shoulders and shoulder blades, which indicated a spinal distortion which curved from side to side, known as first-degree scoliosis. Three (0.5%) students had more pronounced symptoms which indicated second-degree scoliosis. Three hundred and ninety-seven (75.2%) students from urban schools had symmetrical shoulders and shoulder blades and a straight spine, which means they had a good body posture. One hundred and thirteen (21.4%) students had asymmetrical shoulders and shoulder blades, which indicated a spinal distortion which curved from side to side, known as first-degree scoliosis. Eighteen (3.4%) students had more pronounced symptoms which indicated second-degree scoliosis. In the final measuring, 524 (90.8%) students from rural schools have symmetrical shoulders and shoulders blades and a straight spine, which means they had good body posture. Fifty-three (9.2%) students had asymmetrical shoulders and shoulder blades, which indicated a spinal distortion which curved from side to side, known as first-degree scoliosis. However, the extreme deviations or second-degree scoliosis was not present with the students from rural schools in the final measuring. Four hundred and twenty-one (79.7%) students from urban schools had symmetrical shoulders and shoulder blades and a straight spine, which means they had good body posture. One hundred and seven (20.3%) students had asymmetrical shoulders and shoulder blades, which indicated a spinal distortion which curved from side to side, known as first-degree scoliosis. However, the extreme deviations or second-degree scoliosis was not present in the students from the urban schools in the final measuring.

3.1.4 Table 10

After measuring the spine curvature of the students from rural schools, the data obtained showed that, out of 1105 students in the initial measuring, 412 (71.4%) students had symmetrical shoulders and shoulder blades and a straight spine, which means they had good body posture. Ninety (15.6%) students had heads which protruded backward; a flat or bulging chest; a pelvis which was completely shifted forward and down, a bulging, untoned stomach; and hips which shifted forward. All of these symptoms indicated a deformity called first-degree lordosis. Seventy-five (13.0%) students had more pronounced symptoms which indicated second-degree lordosis. Three hundred and five (57.8%) students from urban schools had symmetrical shoulders and shoulder blades and a straight spine, which means they had good body posture. One hundred and thirty (24.6%) students had heads which protruded backward; a flat or bulging chest; a pelvis which was completely shifted forward and down; a bulging, untoned stomach; and hips which shifted forward. All of these symptoms indicated a deformity called first-degree lordosis. Ninety-three (17.6%) students had second-degree lordosis. In the final measuring, 492 (85.3%) students from rural schools had symmetrical shoulders and shoulder blades and a straight spine, which means they had good body posture. Eighty-five (14.7%) students had heads which protruded backward; a flat or bulging chest; a pelvis which was completely shifted forward and down; a bulging, untoned stomach; and hips which are shifted forward. This indicated the deformity, first-degree lordosis. However, the extreme deviation, second-degree lordosis, was not present in the students from the rural schools in the final measuring. Three hundred and ninety-nine (75.6%) students from the urban schools had symmetrical shoulders and shoulder blades and a straight spine, which means they had good body posture. One hundred and twenty-nine (24.4%) students had heads which protruded backward; a flat or bulging chest; a pelvis which was completely shifted forward and down; a bulging, untoned stomach; and hips which shifted forward. This indicated the deformity, first-degree lordosis. However, the extreme deviations, second-degree lordosis, were not present in the students from the rural schools in the final measuring.

4. Citing sources

A decrease in physical activity, due to an urban lifestyle, inadequate exercises during physical education classes, and lack of participation in sports activities, leads to a weakening of the whole muscular system including the muscles within the spine. This leads to specific disorders connected to improper posture and the appearance of physical deformities [15].

Based on current research and statistics, deformities in children's posture appear mostly because of muscle weaknesses in the back area, chest, or abdomen. Thus, a weakness in the pelvic muscles and lower extremities could lead to a secondary disorder in the upper body. Initial changes usually appear first in the muscles and then in the ligaments and skeletal system [16].

Reducing physical activity also reduces the resistance of the locomotor apparatus on the effect of the external factors which enables the appearance and development of many postural disorders [17].

Today, the lifestyle of young people is characterized by a lack of movement, that is, it has become more sedentary. Almost one-third of all children spend more than 4 h a day sitting in front of the TV. This does not account for the time spent sitting in the school or being on the computer and playing video games. All of these tell us that the number of sitting hours, whether at home or in school, with children and youth is growing [18].

All of this is confirmed by research which shows that the percentage of school children and youth with postural disorders is well over 60%. The biggest percentage of these disorders is due to spinal deformities [19, 20]. Unfortunately, the number of children with problems due to body posture is increasing every year [18].

Today, a lot of attention is given to researching the factors that are affecting the appearance of spinal deformity. In some European countries, the law states that the weight of a school backpack must not exceed 10% of the weight of that child [21, 22]. Research conducted worldwide shows that the average weight of a backpack exceeds the permitted amount and ranges from 10 to 14% [21].

Recent researches in some Dalmatian elementary schools, with students from all grade levels, have shown that the average weight of a school bag, with regard to the weight of a student, ranges from 12.5 to 13.8%. A specified, heavier load in a backpack affects the growth and development of the child because the child's locomotor system is not fully developed. In research by different authors around the world and in Croatia, they mention this problem of overly heavy school bags as one of the common factors that are responsible for the development of incorrect physical posture and spinal deformity [23].

Besides the backpack, scientists also mention that school furniture is not being adapted to the growth and development of the children in classroom. This stated that lordosis is a curvature of the spine whose convexity is facing forward. This curvature is present in the cervical and lumbar parts of the spine and has a strong connection with kyphosis [24].

Kyphosis is more present in male students from urban areas, and lordosis is more present in female students than the students from rural areas. A rise in kyphotic diagnoses indicates a sudden growth and development of the skeleton, a lack of strength in the back muscles, an absence of physical hygiene, insufficient physical activity, as well as a lack of the preventive-corrective exercising models [15, 25].

The results of the research have shown that there was a difference in the students from the first to the fourth grade in elementary schools who had presented symptoms related to kyphosis. Two hundred and forty-six (56.5%) students from urban areas, compared to students from rural areas [26].

Out of 299 examinees, 136 (45.48%) examinees were male, and 163 (54.52%) examinees were female. Kyphosis in the male examinees from urban areas was significantly more present (62.02%) than in the case of female examinees (37.98%) [27].

It was determined that, out of 258 examined boys, 142 or 55%, and 111 girls or 43.8% from urban areas, there was an increased thoracic curve (kyphosis) on the spine compared to students from rural areas. The kyphosis chart, depending on the degree of development, focuses on certain characteristics which include mainly the gender of the examinees as well as the type of living environment. It was proven that, in all of the students, the functional disorders were the most present. With the boys, the value is Rf = 73.9 and with the girls Rf = 81.1. Incomplete fixated disorders made up 25.4% of the cases in the boys and 100% of the cases in the girls. The most difficult form or the so-called fixated kyphosis included only one boy (0.4%) [28].

In measuring the postural disorders of the spine, the researchers determined that out of the total number of 320 examinees, 146 students (46%) had the spinal deformity (kyphosis), that is, where the silhouette is hunched forward. This indicates that the children have shoulders which are bent forward, an enhanced stoop, a recessed chest, protruding shoulder blades, a bulged and flabby stomach, and knees which are slightly bent and shifted forward. This indicates the deformity, kyphosis [29].

While measuring the postural disorders of the spine, it was determined that out of the total number of 320 students, 47 students (15%) had asymmetrical shoulders and shoulder blades. This indicates a distortion of the spine which is a sideway curvature, better known as scoliosis [29].

While measuring spine deformities with preschool children, it was proven that the girls from urban areas had the highest deformity percentages: scoliosis (31.25%), lordosis (18.75%), and kyphosis (9.37%). The boys from urban areas had the biggest percentage of deformities: kyphosis (13.33%), scoliosis (31.25%), and lordosis (2.22%) [30].

The results of this research have shown that in the students from the first to the fourth grade of elementary schools, the frequency of the spinal deformity, lordosis, was present in 9 (2.0%) students from urban areas, compared to the students from rural areas [26].

According to the obtained results, lordotic disorders are more present than the kyphotic disorders in the students from urban areas, compared to the students from rural areas. This coincides with this researcher's results. Increased lordotic deviations were recorded in 146 boys (56.6%). One hundred twenty-five girls (49.4%), with even their number of kyphosis, have shown to have slightly better spinal posture than the boys. The greatest number of lordotic deviations refers to the functional disorders or the so-called bad body posture. With the boys from urban areas, their relative frequency is 78% and with girls 77%. With incomplete fixated disorders, it has been found to be in 19.1% of the cases in boys and 23% of the cases in girls. So, with the reducible lordotic disorders, it has been found in 97.1% of the cases in boys from urban areas and 100% in the cases of girls compared to the students from rural areas. The heaviest case, that is, fixated lordosis, was found in four boys from the urban areas, which makes up 2.9% of the total number of disorders. We can conclude from these results that the boys from the urban areas had a slightly larger total percentage of functional disorders. Thus, they showed worse body posture than the girls from rural areas [28].

In measuring the postural disorders of the spine, it was determined that, out of the total number of 320 examinees, 97 students (30%) had the spinal deformity, lordosis. This diagnosis is indicated by a head which protrudes backward; a flat or bulging chest which is a bit lowered; a pelvis which is completely shifted forward and down; a bulged, untoned stomach; and untoned hips which are shifted forward [29].

5. Conclusion

Based on the obtained results of the spinal measurements, that is, before and after the realization of the program, it can be concluded that all of the spinal deformities were more present in the children from urban area than in the children from rural areas. Compared to urban children, everyday responsibilities of the village children included physical activity outside during the day as well as a balanced diet. On the other hand, urban children often ate unhealthy food from the school kitchen, such as pizza, croissants, pies, various pastries, etc.

The students are still not aware enough of their need to choose fruits and vegetables and eat less sweet food. Food with artificial flavors is easily within reach, and once they have tasted it, everything else tastes bad. This, then, negatively affects body posture and is causing deformities in an increasing number of children. It is also known that there are also other health difficulties which can arise because of bad nutrition and lifestyle. Another contributing factor to these disorders is also the fact that urban children have a harder time making new friends. They do not go outside but socialize on Facebook and are on the computer more than ever before.

The research, which compared children from the city to children from the village schools, definitely confirmed that village children have a significantly smaller percent of postural disorders of any kind [12]. Today, school children can have various means which could enhance their quality of life. There are many sport clubs, such as collective and individual sports.

While in urban surroundings there are natural resources and developed and water sports, such as some extreme sports, research in this area shows us that a large percentage of children from elementary schools do not participate in sports activities. Thus, this is the reason for the increased number of children with incorrect posture.

Spinal deformities are present to a great extent because of the way regional landscapes affect a child's spine and the child's way of life. This can be observed in everyday activities that are centered on very little body movement, overeating, and overall inappropriate nutrition. It leads to an unhealthy lifestyle. Thus, there is a need for intervention in order to change the child's behavior and help him to live out a healthier lifestyle. The goal would be to protect and enhance his health through a healthy diet and physical activity.

It is predicted that the number of these children will keep increasing if the significance of the physical activity is not taken seriously enough. Physical education in school can have a great influence on the health of the children. It needs to be understood how great the significance of the games and exercising for the health is and that it is needed by every individual. It has been proven that the frequency of the postural disorders could be significantly decreased with a well-planned corrective procedure which is carried out in the long term. It is necessary to develop an awareness that points to the need for a healthy lifestyle, which includes proper corrective exercises that work to prevent and correct one's posture. Otherwise, the consequences could be devastating not only esthetically but also in a poorer life quality which affects motor capabilities and can endanger one's overall health. That is why there is a need to have a solid, long-lasting program in conjugation with health institutions, which starts from preschool and goes up through high school and focuses on early discovery, prevention, and correction of physical disorders in children and youth. Besides the regular physical education program and sports and dance as electives, a special significance should be placed on the need to organize corrective gymnastics classes. It is very important to point out that only the early detection of the deviations from the correct physical posture is the key to success. Incorrect body postures developed because of weakened postural muscles in the earlier stages of development could be corrected with additional physical exercising programs especially if they are detected earlier on. Early diagnosis is the most important key to a successful treatment of these disorders of the musculoskeletal system. The teachers in physical and health education need to head this off by creating a plan for early detection. The program that was applied in this research had significant effects on the prevention of the postural disorders of the spine.

The applied program has easily fit into the curriculum of the school institutions. We can say that this program and many others, along with certain modifications, could be incorporated into the regular physical education program in order to decrease the number of children with these disorders. When one considers the severity and consequences of these disorders, it can be said that there is still not enough awareness among parents, teachers, and children about the necessity to take certain measures in order to prevent and correct these disorders. However, only an educated teaching staff can contribute to correcting hygiene and exercise which will accomplish the goal of correct body posture. In addition to participating in regular physical and health education classes, students should regularly exercise at home in order to positively affect their growth and development.

Author details

Elvira Nikšić[1]* and Edin Beganović[2]

1 Faculty of Teacher Education, University of Sarajevo, Sarajevo, Bosnia and Herzegovina

2 Faculty of Sports and Physical Education, University of Sarajevo, Sarajevo, Bosnia and Herzegovina

*Address all correspondence to: elvira.beganovic1982@gmail.com

References

[1] Beganović E. Elementary games in the function of improper body posture of students in classroom teaching [Doctor's thesis]. Sarajevo: Faculty of Teacher Education in Sarajevo; 2013

[2] Wickens JS, Kiputh OH. Body mechanic analysis of Yale University freshmen. Research Quarterly. 1937;**8**: 37-48

[3] Stefanović D, Finogenov N, Tasić N, Rašić M, Nikolić D, Jovanović S, et al. The frequency of physical deformities and poor body posture school children in the Republic of Serbia. In: Proceedings of the First Congress of School Doctors Croatian; Split-Trogir; 1972. pp. 351-357

[4] Tribastone R. Compendio Ginnastica Correttiva. Rome: Societa Stampa Sportiva, Roma; 1994

[5] Grimmer K. An investigation of poor cervical resting posture. The Australian Journal of Physiotherapy. 1997;**43**(1): 7-16

[6] Obradović B, Milošević Z. Postural status of preschool children in Novi Sad at the age of 6. Messenger of the Anthropological Society of Serbia. 2008; **45**:375-383

[7] Krsmanović R. Identifying the posture of students in the seventh grade as a condition of the formation of homogeneous groups in order to prevent and eliminate defects. In: Proceedings of the VIII Summer School PFKJ, Šibenik; 1988

[8] Krsmanović T. Postural disorders and how to prevent them. Messenger of the Anthropological Society of Serbia. 2007;**42**:345-351

[9] Ristić V, Marković V, Ljubić M. Bone and joint deformities in primary school students in the municipality of Bojnik. Acta Medica Medianae. 2002;**3**:63-67

[10] Jovović V, Čanjak R. Frequency and structure of the postural disorders with students of younger adolescent age in Montenegro. In: Proceedings Book - Zbornik Radova, Vol. 3; International Scientific Conference; Banja Luka; 2011. p. 33

[11] Živković D, Milenković S, Drobnjak D. The condition of postural disorders and physical deformities of children of younger school age in municipalities of Zaječar, Kruševac and Čačak. Sport Mont. 2004;**2–3**(2):421-426

[12] Živković D, Karaleić S. The presence of bad body posture of younger school aged students based on the kind of sport they play. Physical Education. 2006;**2**: 186-188

[13] Medojević S, Jakšić D. Differences in postural disorders between boys and girls from the age of 7 to 15 in the territory of Vojvodina. In: Collection of Papers of Interdisciplinary Scientific Conference with International Participation "Anthropological Status and Physical Activity of Children, Youth and Adults." Novi Sad: Faculty of Sports and Physical Education; 2007. pp. 49-54 (In Serbian)

[14] Wolansky N. Physical growth and development with monitoring of the body posture. In: Teachers' Manual. Varšava: Polish Medical Publishing House PZWL; 1975

[15] Bogdanović Z. Spinal Deformities in the Saggitudinal Line—Preventing and Correcting. Monography. Novi Pazar: Interprint. State University in Novi Pazar; 2008

[16] Babajak J. Shaping Exercises as an Agent of Preventing the Occurrence of Incorrect Body Posture, Current

Practice No. 1. Novi Sad: Provincial institute for sports; 1984

[17] Jovović V, Čanjak R. The presence of chest and blades disorders of school children of different ages. Sports and Health. 2010;**2**(1):140-146

[18] Lafond D, Descarreaux M, Normand MC, Harison DE. Postural development in school children: A cross-sectional study. Chiropractic & Osteopathy. 2007;**15**:1746-1340

[19] Jovović V. Physical Deformities of Adolescents. Nikšić, Montegraf: Faculty of Philosophy; 1999

[20] Radisavljević M, Koturović L, Aranđelović M. The contribution to the study of shape of the foot of first graders from different socio-economic backgrounds. Physical Education. 1982; **5**:446-469

[21] Fošnarič S, Delčnjak-Smrečnik I. The physical burden of pupils with regard to the weight of their school bags from the eighth year of primary school to ninth year of primary school. Informatologia. 2007;**40**(3):207-210

[22] Gent C, Dols J, Rover C, Sing R, Vet H. The weight of schoolbags and the occurrence of neck, shoulder and back pain in young adolescents. Spine. 2003; **28**(9):16-21. PubMed

[23] Paušić J, Kujundžić H. The weight of the school bag of school aged children. In: Maleš UB, editor. Proceedings of the 3rd International Conference "Contemporary Kinesiology," Mostar. Split: Faculty of Kinesiology; 2008. pp. 183-187

[24] Bjeković G, Čalija M. Incorrect Posture Parents-Educators, Physical Development, Health, Beauty Seems Your Child is in Your Hands. East Sarajevo: Faculty of Physical Education; 2000

[25] Jovović V. The transversal analysis of kyphosis frequency in adolescent students. Messenger of the Anthropological Society of Yugoslavia. 2003;**38**:177-183

[26] Živković M, Bogdanović Z, Midić D, Herodek K, Stojanović S. Postural disorders of school children. Sport Mont, no. 28-30/IX. Podgorica. 2011; **2011**:358-367

[27] Marković Ž. Kyphotic body posture and how owning a computer and gender affect this. Sport Mont, no. 25-27/VIII. Podgorica. 2011;**2011**:366-372

[28] Jovović V. The frequency and structure of sagittal disorders of the spine of the adolescent students. Sport Mont, no. 23-24/VIII. Podgorica. 2010; **2010**:98-103

[29] Beganović E, Bešović M. Presence of the obesity and incorrect body posture in school children. In: Proceedings of the Fifth International Congress "Ecology, Health, Work, Sport." Society "Health for all!" Banja Luka; 2012. pp. 202-210

[30] Hadžikadunić M, Balta S. Correction of body posture with preschool children [Graduate Work]. Sarajevo: Faculty of Sports and Physical Education, University of Sarajevo; 2000. pp. 48-49

Chapter 4

Working Conditions and Health Inequalities

Anna Maria Giammarioli

Abstract

Over the last decades, there has been a considerable progress made to address risks at workplaces and to promote occupational safety and health of workers. Nevertheless, the recent changes of the labor market underline that new risks to the health and well-being of workers should be considered. In this context, a vast amount of studies have analyzed the relationship between work conditions, social inequalities, and health, suggesting a complex net of causation. Only recently, it has been shown that people in lower socioeconomic positions incur higher working risks. The 2008–2013 economic crisis also introduced a reduction of the number of workers in full-time permanent employment with a steady expansion of atypical and precarious workers. The latter have generally been associated with more insecure and unhealthy working conditions. Another important aspect of safety in the workplace is gender differences. Although nowadays there is more information than before about the types of health problems and accidents women incur at the workplace, the gender-related questions are still open issues that require a careful evaluation of work-related risks of men and women. In this chapter, we focused on the current state of the art in the field of occupational health and examined the aspects that are still being debated.

Keywords: occupational health, occupational exposure, risk assessment, social determinants of health

1. Introduction

The target of Occupational Safety and Health (OSH) programs includes fostering a safe and healthy work environment. Successful safety interventions depend on the correct identification of causality mechanisms from exposure to hazards to the onset of disease or injury. The central point of OSH programs is the risk assessment process in which we must identify things, situations, and processes that may cause harm to people. The usual approach to OSH implies three phases: (i) hazard identification related to specific work tasks, (ii) risk assessment or evaluation of the risk associated with that hazard, and (iii) actions and procedures to eliminate the hazard, or control the risk when the hazard cannot be eliminated [1, 2]. It is important to underline the difference between hazards and risks, as many people use the terms interchangeably. Commonly used definitions follow: a hazard is any source of potential damage, harm or adverse health effects on something or someone. A risk is the chance or probability, high or low, that a person will be harmed or experience an adverse health effect if exposed to a hazard [1, 2]. For example, the disease tuberculosis (TB) is the adverse health effect caused by the *Mycobacterium tuberculosis* also defined as the "hazardous biological agent." The risk to get sick from TB depends on the probability to be exposed to the " hazardous biological agent," for example, by working with biological samples infected with *Mycobacterium tuberculosis* or having close contact with someone with the disease TB. In this example, the probability (risk) of being exposed to TB will be very high for people working in the hospital infectious diseases department, but very low for people working in other workplaces, such as a library. Today, there are known many different types of hazards, which can cause adverse effects or harm in the workplace. Hazards can come from a wide range of sources and can be found in every workplace. Workplace risk awareness has grown over time, thus now it is possible to identify situations and processes that are inherently dangerous, such as those associated with chemical, physical, and biological procedures or ergonomic risk factors. Unfortunately, there is no cultural preparation to address risk assessment of new and emerging categories (e.g., the work organization) that do not seem to be inherently dangerous [3–5].

Before going into specific aspects of OSH, it should be emphasized that adverse health outcomes in the medical field can be considered as based on two different approaches: the first considers bad health as an inevitable result of individual behavior patterns; the second considers that poor social and economic circumstances affect health throughout life. These different approaches can be extrapolated into the occupational field. Unfortunately, the occupational safety and health management system has so far given little attention to aspects related to the social-economic organization in which people live and work. In recent years, several studies have shown that safety interventions may be more effective at preventing the incidence of work-related diseases by giving priority to the characteristics of organization structures [6–8]. What does this specifically refer to? Generally, when referring to social-economic organization, we consider some occupational and working conditions, type of organizational structure of companies, different types of contracts (such as atypical jobs), size of the production units, and feminized and masculinized jobs.

In this chapter, assuming that most aspects of common workplace hazards should have been dealt with extensively in other chapters of this book, we will not discuss them unless these can be modified by the specific characteristics of subgroups of workers. Our goal is to provide a point of reflection on the relationship that associates socioeconomic organization and safety interventions in the workplace with social inequalities in health also named health inequalities.

2. Materials and methods

A non-systematic literature review was conducted, based on a selection of current and high-quality articles. Search strategy was set up on the main keywords utilized both in the field of occupational health and health inequalities. The search covered PubMed, Science Direct, and Google Scholar databases. Articles and reports that are considered milestones in the field of health inequalities have also been added to the list of references. Reports from World Health Organization (WHO) and Occupational Safety and Health (OSH) were downloaded from the official websites, and web addresses have been reported in the references. Data relevant to the objectives of this chapter have been synthesized using interpretive analysis.

3. Health inequalities

Health inequalities are well documented in a large number of studies from a broad range of industrialized countries (seen in low-, middle-, and high-income countries) [9–12]. Among the most important scientific evidence of health inequalities, it is necessary to give particular emphasis to the famous Whitehall Studies (I and II) of British Civil Servants led by Michael Marmot [13, 14]. The Whitehall cohort study examined mortality rates of Civil Servants and was conducted over a period of 10 years, beginning in 1967. A long-term follow-up of people enrolled in the two studies is still ongoing [15, 16]. Why the impact of these studies so important? The Whitehall studies concentrate on one "working environment" (British Civil Servants) in which there is little heterogeneity in the social economic position within occupational levels and clear social divisions between levels [17]. Whitehall studies showed that people of lower hierarchical occupational levels had worse health and shorter life expectancy than those who were in higher occupational levels. Whitehall studies likewise showed a gap of 5 years in life expectancy between people at the top and at the bottom of the occupational levels. Whitehall studies have also demonstrated an inverse relationship between social economic position and health as well as mortality related to a wide range of diseases. Based on the results of Whitehall studies, Marmot identified "the social gradient in health" where people are positioned by degrees of affluence and deprivation [13, 18]. People near the top have poorer health than those at the very top but better than people behind them in the health gradient scale. Thus, the social gradient in health means that health inequalities affect everyone in a different manner. In addition to Whitehall research, other studies have provided overwhelming evidence for health inequalities and their distribution in the social gradient of health [19–25]. Health inequalities, within and between different countries, are influenced by an unequal distribution of economic, social, and environmental conditions. People with a lower level of education, a lower occupational class, or a lower level of income tend to die at a younger age and to have a higher prevalence of most types of health problems [11, 17]. It is interesting to note that the health gradient cannot be explained taking into account only the biological or genetic characteristics of people, but rather must be considered as the consequence of the socio-economic conditions in which people live and work. The social gradient in health is a term used to describe the phenomenon whereby people who are less advantaged in terms of socioeconomic position have worse health (and shorter lives) than those who are more advantaged. It is also important to highlight that social gradient of health considers differences between social groups rather than between individuals.

4. What are the factors that influence people's health status?

Scientific work on health inequalities has exponentially increased over the last five decades and particularly since the establishment of the World Health Organization (WHO)'s Commission on Social Determinants of Health (CSDH) in 2005. The CSDH approach has focused on the Social Determinants of Health (SDH) perspective, providing an alternative for the approaches limited only to the medical-health aspect and individual behaviors [26]. As a matter of fact, the medical-health approach had always focused on improving health care quality and addressing unhealthy behaviors (e.g., incorrect life style) to achieve greater health equality. Thus, individuals have been considered as responsible for their own health, and the main strategy for preventing disease has been focused on the promotion of correct life style and on behavioral modifications (e.g., smoking cessation, decreasing salt and fat intake, and reducing sedentary lifestyle) [21, 27]. Encouraging better individual behavior is a well-established approach to health promotion, but the evidence suggests that these interventions may have limited effect without to tackle health inequalities [21]. Conversely, an incorrect lifestyle could be a response to social breakdown and a mirage to escape from social adversity and stress. Several studies have shown that alcohol dependence, illicit drug use, and cigarette smoking are all closely associated with markers of social and economic disadvantage [28, 29]. The WHO introduced a new approach for public health intervention that recommended more concern toward social policies and social determinants of health. According to the suggestions of the WHO [10, 30], different countries in the world (especially European countries) are focusing their health policy interventions both on promoting better lifestyles and addressing the root causes of health inequalities.

4.1 Social determinants of health

Social determinants of health include all the major non-genetic and non-biological factors that influence human health. In other words, they are the socioeconomic factors operating in the society that ultimately lead to poor health outcomes. The field of the social determinants of health is perhaps the most complex and challenging of all [9–12]. As reported on the official website of the WHO [9], social determinants of health are defined as follows:

"The conditions in which people are born, grow, work, live, and age, and the wider set of forces and systems shaping the conditions of daily life. These forces and systems include economic policies and systems, development agendas, social norms, social policies and political systems" [9].

Several "conceptual models" have been developed to describe the complex process by which social hierarchies are associated to health. The models try to illustrate how health inequalities are created through the effects of social stratification [11, 31, 32]. Generally, the classifying of individuals into groups with different relative social positions is based on characteristics such as education, income, labor market position, ethnicity, and gender. All together, they are also named social determinant of health. The uneven distribution of opportunities and resources (coupled with social positions) is associated to systematic differences in living conditions and to differential vulnerability. It is important to emphasize that although the "conceptual models" differ in style and complexity, most of these represent health as the outcome of a chain of events or social influences, including both proximal and distal determinants of health. Proximal factors act directly or almost directly to cause disease, (e.g., individual lifestyle factors, housing, water and sanitation, and social and community networks), and distal determinants act indirectly (including social resources like education, employment opportunities, political influence, income, and property). The most significant distal factor is the social structure of society, which is the upstream in the causal chain and acts via a number of intermediate causes establishing person's position in the social hierarchies. Numerous studies have shown that a "dose-response" association with health [17, 26] characterizes education and income. As stated by the WHO and by experts in the field of health inequalities, six main areas of interventions are identified for good health condition:

1. Give every child the best start in life.

2. Enable all children, young people, and adults to maximize their capabilities and have control over their lives.

3. Create fair employment and good work for all.

4. Ensure a healthy standard of living for all.

5. Create and develop healthy and sustainable places and communities.

6. Strengthen the role and impact of ill-health prevention.

In the last few years, new and stronger scientific evidence and several conceptual models on the levels of causation of health inequalities have been developed [31, 32]. A detailed description of the six areas as well as deeper conceptual models on the levels of causation (by which social hierarchies are created) goes beyond the scope of the present chapter and interested readers can refer to more focused reviews [9, 11, 14, 26, 31, 32]. Here, we like to emphasize that the importance of employment and the quality of work is recognized as one of the most important key determinants of health.

5. Employment and working conditions

To analyze how the working world can affect the health of populations, we need first to clarify three occupational aspects: employment relations, employment conditions, and working conditions. The former concerns the various relationships between the employer and employee. In the developed countries, "employment relations" are often subject to the enforcement of specific law or a contract of hire. In the developing and under-developing countries, most employment agreements are not explicitly subject to any formal law or contract. Considering that the "employment relations" varies both within and between countries, the term "employment conditions" was introduced by the WHO [33], classifying different types of employment into six "dimensions" of global scope:

1. Employment

2. Precarious employment

3. Unemployment

4. Informal employment and informal jobs

5. Child labor

6. Slavery/bonded labor

Finally, "working conditions" are related to the tasks carried out by workers, the physical and chemical work environment, ergonomics, the way the work is organized, the social work environment, and the technology being used. Several occupational studies have shown that the relationship between employment condition and work conditions reflect and reinforce the social gradient of health [22]. Thus, people with lower education and lower occupational positions have been associated with dangerous work and with worse work-related exposure [13, 17, 23, 24]. Generally, they were subjected to poorer health than those in higher positions [23, 24]. However, the inequality of exposure does not seem to be sufficiently recognized by all countries in the world. For example, in developing countries, there is a great lack of research, as well as data about risk factors and exposure levels for specific jobs [25]. Nevertheless, in this section, we will try to identify common features and trends among and between economically different countries. Our analysis on occupational conditions starts with the most disadvantaged "dimensions" as registered in the list of six "dimension" of global scope. The most disadvantage conditions are slavery/bonded labor and child labor, as these conditions are often not in line with the application of human rights.

5.1 Slavery/bonded labor

After abolition of slavery practice, which occurred at different times in different countries, its legacy still persists and influences health outcomes. An estimated over 24.9 million people were victims of modern slavery or forced labor in 2016 [34]. Half of these were in debt bondage, in which personal debt is used to forcibly obtain labor. Currently, these numbers do not seem to have decreased. (Data according to a collaborative research performed in 2016 by the International Labour Organization (ILO) and the Walk Free Foundation in partnership with the International Organization for Migrant (IOM) and benefiting from inputs provided by United Nations Office (ONU) of the High Commissioner for Human Rights (OHCHR) [34].) Although modern slavery is not defined in law, it is used as an umbrella term that focuses attention on commonalities across some legal concepts. The term "slavery" includes different forms of "illegitimate work" (e.g., bonded labor, slavery by descent, the worst forms of child labor, and any type of human trafficking). Slaves are forced to work for little or no pay and under threat or coercion.

Slaves can be found in its various forms, practically in all countries and all economic activities [34]. Slaves live in vulnerable conditions [35] in the lowest level of socio-economic gradient.

To study the relationship between slaves and health is very complex due to clandestine nature of duties associated and the denial of the authority regarding its existence. Studies on slavery have mainly given a qualitative picture of disease patterns. Slaves suffer physical and mental trauma as well as of malnutrition due to coercive action, but also restriction of movement and violence. Slaves are often constrained in poor working conditions and are exposed to more hazardous and adverse conditions than other workers. Generally, they are reluctant to access health and social services following accidents or ill health. Some evidence has shown on the adverse health outcomes and health inequalities as a result of physical violence and mental trauma, risky behavior, absence or inaccessible welfare measures, and cultural barriers.

5.2 Child labor

Child labor is a violation of fundamental human rights. Based on data of UNICEF, ILO, and World Bank, an estimated over 168 million children aged 5–17 are engaged in child labor [36, 37]. However, there is no consensus about the definition of child labor.

The Child and Adolescent Labour Prohibition and Regulation (CLPR) Act of 1986 defines "Child" any person below the age of 14 and prohibits employment of a child in any employment including as a domestic help (except helping own family in non-hazardous occupations). According to UNICEF (2006), child labor means children below 12 years of age with the exception of those from 12 to 14 years of age engaged in "light work." For ILO, child labor (regardless of the age of the child) is a violation of fundamental human rights that hinder children's development and potentially leading to lifelong physical or psychological damage [36, 37].

Child labor is the combined product of many factors, such as social discrimination, poverty, and migration. Child labor reinforces intergenerational cycles of poverty by keeping the children of the poor out of school and limiting their prospects for upward movement in the social hierarchy. Child labor is a major cause of illiteracy, low education, and low-skilled workers. As described above, education is a powerful social determinant that creates and strengthens human capital in a society. Every child forced to drop out of school to work is a loss of human capital. This lowering of human capital has been linked to slow economic growth and to poor social development. The ILO study has shown that eliminating child labor in developing economies could generate economic benefits nearly seven times greater than the costs that would need to be invested in better schooling and social services [36]. A growing number of studies have shown that health problems are one of the main negative effects of child labor [35, 38]. With regard to biological differences (in anatomy, physiology, biochemistry, and toxicology), children may suffer from greater vulnerability than adults to hazardous workplace factors. Children have an immature immune system that increases vulnerability to biological agents. Extreme workloads may lead to various health disorders because of children's lesser bone elasticity, strength, and capacity to support heavy workloads. These factors can lead to musculoskeletal symptoms among child laborers [38], but some of these health effects could appear later during adulthood. Ergonomic risks can also result from inadequate dimensions of tools and equipment that are calibrated on an adult shape. Child labor has been associated with problems related to the physical, physiological, mental, and social development of children [35]. Child labor spans various sectors, including agriculture, quarrying manufacturing, mining, and domestic service. In its most extreme forms, child labor involves children being enslaved. Many working children are also involved in other unacceptable work conditions such as war combats, prostitutions, and drug selling.

5.3 Informal employment and informal jobs

These are non-regulated placements in the labor market, which usually involve an informal arrangement between the employee and employer (informal employment) or self-employment (informal jobs). The latter implies exchange of products or services rather than exchange of labor force. In several countries, workers' rights are related to a formal job contracts. Thus, the workers' entitlement for social benefits, such as paid retirement, sick or maternity leave, or access to health care, requires a formal job contract not covered by informal employment or informal jobs. Besides a lack of social benefits, workers holding informal employment or informal jobs have lower salaries, high turnover, lack of security, non-defined work-time, lack of compensation at firing, and limited unionization [39, 40]. Relations between informal economy/informal jobs and health are not often studied and health inequalities even less. Overall, employment status and other occupational data are not always available. Most of the available studies are qualitative case studies or descriptive surveys carried out on informal employees versus formally hired workers. The qualitative methods are more descriptive and focused on the description of individual experiences [39, 41, 42]. Thus, the lack of official numerical data penalizes the application of quantitative statistical methods. The studies based on workers in the informal economy are penalized by the large heterogeneity of occupations and trades, private work agreements, and the lack of risk assessment due to uniqueness of workplaces and the scattered spatial distribution of workers. Informal workers appeared more vulnerable to adverse effects of health than those engaged in formal jobs [43]. Informal employment may cause mental distress and psychological diseases, because of job insecurity, or the fear to lose long-term jobs. Job insecurity is always associated with specific adverse health effects related to mental distress and psychological diseases [44, 45]. A positive association has also been shown between informal jobs and years of life lost in disability or due to early death for all diseases. These studies, however, have not considered the heterogeneity of the informal sector that includes small entrepreneurs, self-employed, and salaried workers. In addition, few studies have described the relations between informal employment and health inequalities, describing mainly health effects without considering the social position of the subjects examined (probably a small entrepreneur is located in a different socioeconomic position respect to an informal employee). Specific studies will be needed to establish a correct relationship between each specific informal employment with health and health inequalities.

5.4 Unemployment and precarious employment

In 2008, the world economy faced its most dangerous crisis since the Great Depression of the 1930s. The so-called "subprime mortgage" crisis began in the United States in 2007 and spread quickly, first to the entire United States financial sector and then to financial markets of Europe. Changes that took place in the economy have influenced the labor market, increasing unemployment and the number of precarious workers. The burden of the recession has not been equally distributed among the different socio-economic groups. People in lower socio-economic positions have had a higher risk of dismissal, unemployment, and precarious employment, thus strengthening the social gradient of health.

For example, workers with weaker employment contracts, less skilled, and less experienced workers, as well as those from some ethnic minority groups have suffered the greatest load of the recession. Since crisis, the economy of industrialized countries has continued to be slowed, but with steady recovery. However, the growth of jobs has been predominantly in higher skilled employment, while the number of manufacturing and low-skilled jobs has been declined over a longer period. Many workers have been trapped in a cycle of low-paid, poor quality jobs (many of which are precarious jobs), and unemployment. In this section, the conditions of unemployed people and precarious workers will be considered together as they share insecurity, low self-esteem, and lack of control over work and home life and stressful circumstances that can cause adverse health effects such as anxiety, mental effects, and depression [46–48]. Psychosocial risks can increase the chances of premature death [17]. These occur because in emergencies, human hormones and nervous system are ready to deal with an immediate physical threat: raising the heart rate, mobilizing stored energy, diverting blood to muscles, and increasing alertness. For brief periods, this does not matter, but if the tension goes on for too long, people become more vulnerable to a wide range of conditions including infections, diabetes, high blood pressure, heart attack, stroke, and depression. Some studies have emphasized that negative effects are proportional to the duration of unemployment, which progressively damage health [47–49], and negative effects are greatest among those who experience long-term unemployment [50]. They have also increased rates of limiting long-term illness [51], mental illness [46], and cardiovascular disease [52]. There are three ways in which unemployment affects levels of morbidity and mortality: (1) financial problems result in lower living standards, which may in turn reduce social integration and lower self-esteem [47]; (2) unemployment can trigger distress, anxiety, and depression [49] not only among the unemployed themselves but also among their partners and children [53]; and (3) unemployment can impact on health behavior, being associated with increased smoking and alcohol consumption and decreased physical exercise [47]. It is interesting to note that the relationship linking unemployment and poor health runs in two directions. Unemployment contributes to ill health and in turn, poor health increases the likelihood of unemployment, and the two can become mutually reinforcing [54]. The extent to which ill health and disability act as an obstruction to work is highly dependent on educational qualifications. Evidence suggests that health risks are even higher depending on geographical regions where unemployment and work insecurity are widespread [44, 45]. When in work, the long-term unemployed are more often engaged in precarious work. Precarious jobs are also common among young workers, new workers, and adult populations with low skill level and low educational degree. Precarious workers are disproportionately employed in physically demanding or hazardous jobs. This puts them at a higher risk of workplace injuries and illnesses. In addition, in some cases, employers are reluctant to commit economic resources to train unstable workers who will leave their jobs quickly. Thus, precarious workers, suffering from a lack of "information" with regard to safety at work, could be more vulnerable than permanently people to adverse working conditions. From what has been explained so far, getting people into fair employment is an important strategy for improving health.

5.5 Employment

It has been repeatedly emphasized that permanent employees work in safer conditions than the people working in the other employment dimensions. Nevertheless, over the few last years, the global workforce is constantly changed in relation to age and gender. Thus, it is important that the implications of demographic change in relation to OSH are taken into account to help build effective policies and strategies for all workers.

For example, women are entering the workforce in increasing numbers and have a higher risk than men for some work-related disorders (e.g., musculoskeletal disorders). It is well documented that injury rates are significantly higher among young workers than among older or more experienced workers [37–55]. Nevertheless, older workers need adaptive practices and equipment to work safely [56].

The gender sensitive approach has showed a steady growth, while few studies have been performed on age differences. Due to the increased participation of women in the workforce, the European Agency for Safety and Health at Work (EU-OSHA) has encouraged a policy of gender equality in all European member states [57, 58].

5.5.1 Occupational gender differences

The gender-sensitive approach includes "sex" and "gender" characteristics, where "gender" refers to those characteristics of women and men that are socially and culturally determined, whereas "sex" refers to biological and physiological differences. Sex (biology) and gender (the social construction of masculinity and femininity) interact constantly and can lead to gender disparities in human diseases in terms of incidence, prognosis, and response to therapy [59–61]. Gender disparities run in a transverse manner among different social groups along a social gradient of health, and the gender bias according to occupational area is closely related to social inequalities.

Occupational studies should take into account gender differences related to two aspects: (1) socioeconomic and cultural factors and (2) biological factors that change the effect of an equal occupational hazard exposure in male and in female.

Socioeconomic and cultural factors, as well as gender stereotypes, have affected occupational segregation, which is the underlying reason for so many gender inequalities [62]. Women and men have been restricted in "feminized and masculinized" sectors of activity (horizontal segregation) due to gender stereotypes. Women and men carry out the same jobs, but perform different tasks [63]. In addition, men are more likely to work in jobs higher up in the occupational hierarchy than women (vertical segregation) [60, 61]. Hence, gender segregation strongly contributes to an unequal distribution of working conditions as well as exposure to different physical and psychological risks between sexes [64]. Women are also more likely to have part-time or temporary contracts than men [65]. Job segregation strongly contributes to different hazard exposure and consequently to different health outcomes. Examples of these could include skin diseases that women suffer when working with wet hands, chemical cleaning, and sterilizing agents as well as protective gloves containing latex dust. Several studies have also reported male-female differences in the prevalence of symptoms of work-related musculoskeletal disorders, some arising from workplace exposure differences [66]. In this regard, it is important to emphasize that gender-related biological differences may result in differential vulnerability of women and men at the same physical, chemical, and biological workplace factors such as hazardous substances and biological agents [62, 67, 68].

Many studies, on which occupational safety and health are based on, have been performed on men excluding women [60]. In addition, gender differences have been rarely studied in epidemiologic research related to occupational safety and health. For many years, Food and Drug Administration guidelines specifically precluded participation of women in many toxicological studies [69]. Currently, it is believed that women and men differ in many aspects of biological vulnerability to occupational hazards.

Chemicals may induce variable toxic actions according to the amount absorbed by the body (or Body Burden) [70]. Toxicological studies have defined the threshold limit value (TLV) or daily level to which a worker can be exposed without adverse health effects. However, TLV has been calculated on men, and few studies have measured exposure for men and women in the same occupational setting. Toxicity varies depending how quickly and efficiently toxic agent is metabolized. Generally, the relationship between exposure dose, absorbed dose, and effective dose is complex and depending on various factors that are studied by pharmacokinetics or toxicokinetic that is split up in four different phases [(1) absorption across the body barriers (e.g., skin and hair), (2) the distribution into the body, (3) the metabolism, and (4) the excretion] are all subject to sex differences [71]. Absorption probably differs between women and men due to the condition of the skin (with or without cosmetics), number of hair follicles, breathing rates, and respiratory volume. It has been identified a number of chemical agents for which the body burden is different in women and men even when they carry out the same job [70]. Differences in "Body Burden" are clarified by the anthropometric differences between sexes according to muscle mass, fatty tissue, and bone mass. Women have a higher percentage of body fat than men. Adipose tissue makes women more susceptible to dangerous fat-soluble substances, such as organic solvents (e.g., benzene and trichloroethylene). All together these factors affect the extent of distribution of the chemical into the body compartments. However, the greater role in toxicokinetic variability is played by differences in xenobiotic metabolism [71]. Primarily, these reflect the differences in gene expression for enzymes of the CYP450 superfamily, the major family of enzymes involved in the metabolism of chemical agents. Sex-based variance in the expression and activity of CYP isoenzymes are been reported in different studies [71]. In addition, CYP450 activity is also modulated by sex hormones [72, 73]. Finally, renal excretion of compounds is higher in men than in women [71]. It is clear that the limits defined by TLV should be monitored according to gender differences, in order to determine appropriate procedures in OSH.

Work-related asthma is one of the most frequently reported occupational lung diseases, and sensitivity to asthmatic attacks is an emblematic example of gender differences in workplaces as the asthmatic attacks are closely related to hormonal changes [72, 73]. Men are highly sensitive in prepubescent age, but their susceptibility decreases drastically in adulthood. Conversely, women become sensitive to asthmatic attacks in post pubertal age and this condition continues until menopause. Unfortunately, the period of greater sensitive is overlapped with the working age period. These differences could make women more susceptible than men to occupational asthma and indicate the need for additional prevention measures for women (e.g., particular PPE).

6. Work-related stress

Work-related stress can be defined as the adverse reaction people have to excessive pressures or extra demand placed on them at workplace [74]. Generally, work-related stress is due to the type of work, position in the social hierarchy, horizontal and vertical discrimination, sexual harassment, and the situation outside of work. In this scenario, work-related stress is an important intermediate factor linking workers in more disadvantaged socioeconomic positions with poor health [74, 75]. Factors like skills and experience, age, gender, ethnicity or disability may all affect whether an employee can cope. Employees feel stress when they cannot cope with pressures and other issues [74, 75]. Both women and men report high levels of work-related stress but stress affects people differently and what stresses one person may not affect another depending on the context in which people live and work [74, 75]. Stress is not an illness but it can make workers ill. As described above, stress raises blood pressure, increases the risk of heart disease [76], and weakens the immune system. It can cause depression and even lead to suicide and cause a number of mental and physical disorders [77, 78]. Finally, stress can increase drinking or smoking, reinforcing inequalities [28, 29].

7. Discussion

Health inequalities can be defined as differences in health status between different population groups, where people who are less advantaged in terms of socioeconomic position have worse health (and shorter lives) than those who are more advantaged [9, 13, 17, 21, 26, 33]. Until a few years ago, the approaches to resolve health inequalities focused on improving health care quality and addressing unhealthy behaviors (incorrect life styles) to achieve greater health equity [10, 27, 30]. These approaches assume that people are the only ones responsible for their health, when they adopt an unhealthy lifestyle. Similarly, occupational safety and health rules are based on actions and procedures that workers have to address in their specific workplaces [1–5]. Although incorrect behaviors must be avoided, addressing work-related risks as though they exist solely within a workplace is an ineffective and incomplete strategy. Due to the recent social and economic changes, it has become more important than ever to anticipate new and emerging work-related safety and health risks [33, 78].

The world of work has undergone profound transformations related to both demographic changes (e.g., age, gender, and ethnicity) and employment conditions (increasing the number of precarious and informal workers) [33, 78]. Similarly, changes that have taken place in the economy have increased the economic inequalities driven by neo-liberalism. For example, some changes in the organization of work have brought flexibility that allows more people to enter the labor force, but may also lead to psychosocial issues (e.g., insecurity), inadequate OSH, and excessive work hours [33, 78]. Lower level occupational roles and poor working conditions have been more common among people with a lower level of education and lower position in the social gradient [22–25, 33]. Low occupational-skill jobs have often been associated with dangerous work and with worse work-related exposure, reflecting inequalities in exposure to risks [22, 25, 33]. Over the last few years, it has also been established that some organizational aspects can also induce new risks for workers [16, 21, 40, 44, 45, 48, 76, 77]. Among these, psychosocial risks need additional attention in terms of situations and employment practices that affect work-related stress and mental health outcomes [42, 44–46, 48, 76, 77]. In this scenario, people who are discriminated against at work according to age, gender, ethnicity, and contracts of hire should be considered more vulnerable workers than others [56–62, 64, 65, 78].

Finally, it should be considered that occupational accidents and work-related diseases have a substantial global impact, and this impact varies according to where workers live and work. Indeed, it is known that occupational mortality and morbidity are not equally distributed across the world. About two-third (65%) of global work-related mortality is estimated to occur in Asia, followed by Africa (11.8%), Europe (11.7%), America (10.9%), and Oceania (0.6%). This reflects the distribution of both the world's working population and differing levels of national economic development, as well as hazardous work.

8. Conclusion

As described in this chapter, some workers are more exposed to adverse health work environments due to strong and persistent social inequalities. Thus, occupational safety and health interventions could be more effective by giving priority to a holistic view of the hazards that workers experience and the range of adverse effects that occur as a result. Today, prevalent occupational safety and health practices remain disproportionately focused on individual-level with inadequate attention to organizational and higher-level systems such as economic and labor market structures, education and training policy, unionization, and other macro-level policies. In the light of the above considerations, further efforts will be needed to apply a new approach to occupational safety and health and to achieve better health status for all workers.

Conflict of interest

No conflicts of interest were declared.

Author details

Anna Maria Giammarioli
National Center for Global Health, Istituto Superiore di Sanità, Rome, Italy

*Address all correspondence to: anna.giammarioli@iss.it

References

[1] Alert and sentinel approaches for the identification of work-related diseases in the EU (EU-OSHA). 2018. Available from: https://osha.europa.eu/en/tools-and-publications/publications/alert-and-sentinel-approaches-identification-work-related/view

[2] Occupational Health and Safety—Hazard Identification and Elimination and Risk Assessment and Control-Can/Csa-Z1002-12 (R2017) series of occupational health and safety management Standards. Available from: https://www.ccohs.ca/oshanswers/hsprograms/risk_assessment.html

[3] EU-OSHA. Expert Forecast on Emerging Chemical Risks Related to Occupational Safety and Health. Luxembourg: Office for Official Publications of the European Communities. 2009. Available from: https://osha.europa.eu/en/publications/reports/TE3008390ENC_chemical_risks

[4] EU-OSHA. Exploratory Survey of Occupational Exposure Limits (OELs) for Carcinogens, Mutagens and Reprotoxic Substances (CMRs) at EU Member States Level. Luxembourg: Publications Office of the European Union; 2009. Available from: https://osha.europa.eu/en/publications/annual_report/2009full

[5] EU-OSHA. OSH in Figures: Work-Related Musculoskeletal Disorders in the EU—Facts and Figures. Luxembourg: Publications Office of the European Union; 2010. Available from: https://osha.europa.eu/en/publications/osh-figures-work-related-musculoskeletal-disorders-eu-facts-and-figures

[6] LaMontagne AD, Keegel T, Louie A, Ostry A, Landsbergis P. A systematic review of the Jobstress intervention evaluation literature, 1990-2005. International Journal of Occupational and Environmental Health. 2007;**13**(3):268-280

[7] Halperin WE. The role of surveillance in the hierarchy of prevention. American Journal of Industrial Medicine. 1996;**29**(4):321-323

[8] Stewart J. Occupational hygiene: Control of exposures through intervention. In: Aakrog A, Stellman J, editors. Encyclopedia of Occupational Health and Safety. Geneva: International Labor Organization; 2011

[9] World Health Organization. What are the social determinants of health? 2008. Available from: https://www.who.int/social_determinants/sdh_definition/en/

[10] Kurup ES, A & World Health Organization. Equity, Social Determinants and Public Health Programmes/Editors Erik Blas and Anand Sivasankara Kurup. Geneva: World Health Organization; 2010. Available from: https://apps.who.int/iris/handle/10665/44289

[11] WHO. Commission on social determinants of health. A conceptual framework for action on the social determinants of Health. Geneva. 2007. Available from: http://www.who.int/social_determinants/resources/csdh_framework_action_05_07.pdf

[12] Friel S, Marmot MG. Action on the social determinants of health and health inequities goes global. Annual Review of Public Health. 2011;**32**:225-236. DOI: 10.1146/annurev-publhealth-031210-101220

[13] Marmot MG, Smith GD, Stansfeld S, Patel C, North F, Head J, et al. Health inequalities among British civil servants: The Whitehall II study. Lancet. 1991;**337**:1387-1393

[14] Marmot MG. Social differentials in health within and between populations. Daedulus. 1994;**123**:197-216

[15] North F, Syme SL, Feeney A, Head J, Shipley MJ, Marmot MG. Explaining socioeconomic differences in sickness absence: The Whitehall II study. British Medical Journal. 1993;**306**:361-366

[16] Bosma H, Marmot MG, Hemingway H, Nicholson A, Brunner EJ, Stansfeld S. Low job control and risk of coronary heart disease in the Whitehall II (prospective cohort) study. British Medical Journal. 1997;**314**:558-565

[17] Marmot M, Bell R. Fair society, healthy lives. Public Health. 2012;**126**(Suppl 1):S4-S10

[18] Marmot MG, Mcdowall ME. Mortality decline and widening social inequalities. Lancet. 1986;**328**:274-276

[19] Feinglass J, Rydzewski N, Yang A. The socioeconomic gradient in all-cause mortality for women with breast cancer: Findings from the 1998 to 2006 National Cancer Data Base with follow-up through 2011. Annals of Epidemiology. 2015;**25**(8):549-555. DOI: 10.1016/j.annepidem.2015.02.006

[20] Marmot M et al. Lancet. 2005;**365**(9464):1099-1104

[21] Stringhini S, Carmeli C, Jokela M, Avendaño M, Muennig P, Guida F, et al. LIFEPATH consortium . Socioeconomic status and the 25 × 25 risk factors as determinants of premature mortality: A multicohort study and meta-analysis of 1·7 million men and women. Lancet. 2017;**389**(10075):1229-1237. DOI: 10.1016/S0140-6736(16)32380-7. Erratum in: Lancet. 2017;389(10075):1194

[22] Hämmig O, Bauer GF. The social gradient in work and health: A cross-sectional study exploring the relationship between working conditions and health inequalities. BMC Public Health. 2013;**13**:1170. DOI: 10.1186/1471-2458-13-1170.

[23] Schnall PL, Dobson M, Rosskam E, editors. Unhealthy Work: Causes, Consequences, Cures. Amityville NY: Baywood Publishing Co; 2009

[24] Siegrist J, Theorell T. Socioeconomic position and health: The role of work and employment. In: Siegrist J, Marmot M, editors. Social Inequalities in Health. New Evidence and Policy Implications. Oxford: Oxford University Press; 2006. DOI: 10.1093/acprof:oso/9780198568162.001.0001

[25] Montano D, Hoven H, Siegrist J. A meta-analysis of health effects of randomized controlled worksite interventions: Does social stratification matter? Scandinavian Journal of Work, Environment & Health. 2014;**40**(3):230-234. DOI: 10.5271/sjweh.3412

[26] Commission on Social Determinants of Health. Closing the Gap in a Generation: Health Equity through Action on the Social Determinants of Health. Final Report of the Commission on the Social Determinants of Health. Geneva: World Health Organization; 2008

[27] Piccinelli C, Carnà P, Stringhini S, Sebastiani G, Demaria M, Marra M, et al. The contribution of behavioural and metabolic risk factors to socioeconomic inequalities in mortality: The Italian longitudinal study. International Journal of Public Health. 2018;**63**(3):325-335. DOI: 10.1007/s00038-018-1076-8

[28] Marmot M. Inequality, deprivation and alchohol use. Addiction. 1997;**92**:S13-S20

[29] Greenhalgh EM, Scollo MM, Pearce M. 9.7 explanations

of socio-economic disparities in smoking. In: Scollo MM, Winstanley MH, editors. Tobacco in Australia: Facts and Issues. Melbourne: Cancer Council Victoria; 2016. Available from: https://www.tobaccoinaustralia.org.au/chapter-9-disadvantage/9

[30] WHO. Creating Healthy Cities in the 21st Century. Geneva: WHO; 1996

[31] Diderichsen F, Andersen I, Manuel C, Working Group of Danish Review on Social Determinants of Health, Andersen AM, Bach E, et al. Health inequality—Determinants and policies. Scandinavian Journal of Public Health. 2012;**40**(8 Suppl):12-105. DOI: 10.1177/1403494812457734

[32] Hosseini Shokouh SM, Arab M, Emamgholipour S, Rashidian A, Montazeri A, Zaboli R. Conceptual models of social determinants of health: A narrative review. Iranian Journal of Public Health. 2017;**46**(4):435-446

[33] Quinlan M, Muntaner C, Solar O, Vergara M, Eijkemans G, Santana V, et al. Policies and interventions on employment relations and health inequalities. International Journal of Health Services. 2010;**40**(2):297-307

[34] Global estimates of modern slavery: Forced labour and forced marriage, prepared jointly by the International Labour Organization (ILO) and Walk Free Foundation, in partnership with IOM, the UN Migration Agency. Available from: www.alliance87.org/2017ge

[35] Fassa AG. Health Benefits of Eliminating Child Labour. ILO/IPEC Working Paper. International Labour Office, International Programme on the Elimination of Child Labour; 2003

[36] ILO. Investing in Every Child: An Economic Study of the Costs and Benefits of Eliminating Child Labour. Geneva: IPEC; 2004. pp. 4-5. Available from: http://white.lim.ilo.org/ipec/documentos/costos_y_beneficios_eliminacion_ti_oit_2004.pdf

[37] Global estimates of child labour: Results and trends, 2012-2016, prepared by the International Labour Organization (ILO). Available from: https://unitedkingdom.iom.int/blog/news/ilo-walk-free-and-iom-release-the-new-global-estimates

[38] Huk-Wieliczuk E. Physical work load and state of health of school-aged children in the southern Podlasie region. Annals of Agricultural and Environmental Medicine. 2005;**12**(1):95-100

[39] Santana VS, Loomis D. Informal jobs and non-fatal occupational injuries. The Annals of Occupational Hygiene. 2004;**48**(2):147-157

[40] Julià M, Tarafa G, O'Campo P, Muntaner C, Jódar P, Benach J. Informal employment in high-income countries for a health inequalities research: A scoping review. Work. 2015;**53**(2):347-356. DOI: 10.3233/WOR-152176

[41] Bisgrove EZ, Popkin BM. Does women's work improve their nutrition: Evidenc from the urban Philippines. Social Science & Medicine. 1996;**43**(10):1475-1488

[42] Santana VS, Loomis D, Newman B, Harlow SD. Informal jobs: Another occupational hazard for women's mental health? International Journal of Epidemiology. 1997;**26**(6):1236-1242

[43] Parrón T, Hernández AF, Pla A, Villanueva E. Clinical and biochemical changes in greenhouse sprayers chronically exposed to pesticides. Human & Experimental Toxicology. 1996;**15**(12):957-963

[44] Sverke M, Hellgren J, Naswall K. No security: A meta-analysis and review of job insecurity and its consequences.

Journal of Occupational Health Psychology. 2002;7(3):242-264

[45] Ferrie JE, Head J, Shipley MJ, Vahtera J, Marmot MG, Kivimaki M. Injustice at work and incidence of psychiatric morbidity: The Whitehall II study. Occupational and Environmental Medicine. 2006;**63**(7):443-450

[46] Thomas C, Benzeval M, Stansfeld S. Employment transitions and mental health: An analysis from the British household panel survey. Journal of Epidemiology and Community Health. 2005;**59**:243-249

[47] Maier R, Egger A, Barth A, Winker R, Osterode W, Kundi M, et al. Effects of short- and long-term unemployment on physical work capacity and on serum cortisol. International Archives of Occupational and Environmental Health. 2006;**79**(3):193-198

[48] Hämäläinen J, Poikolainen K, Isometsä E, Kaprio J, Heikkinen M, Lindeman S, et al. Major depressive episode related to long unemployment and frequent alcohol intoxication. Nordic Journal of Psychiatry. 2005;**59**(6):486-491

[49] Voss M, Nylén L, Floderus B, Diderichsen F, Terry PD. Unemployment and early cause- specific mortality: A study based on the Swedish twin registry. American Journal of Public Health. 2004;**94**(12):2155-2161

[50] Bethune A. Unemployment and mortality. In: Drever F, Whitehead M, editors. Health Inequalities: Decennial Supplement, ONS Series DS No. 15. London: The Stationery Office; 1997. pp. 156-167

[51] Bartley M. Health Inequality: An Introduction to Theories, Concepts and Methods. Cambridge: Polity; 2004. p. 224. ISBN:0745627803

[52] Gallo W, Teng H, Falba T, Kasl S, Krumholz H, Bradley E. The impact of late career job loss on myocardial infarction and stroke: A 10 year follow up using the health and retirement survey. Occupational and Environmental Medicine. 2006;**63**:683-687

[53] Bartley M, Ferie J, Montgomery SM. Health and labour market disadvantage: Unemployment, nonemployment and job insecurity. In: Marmot M, Wilkinson RG, editors. Social Determinants of Health. 2nd ed. Oxford: Oxford University Press; 2006

[54] Schuring M, Burdorf L, Kunst A, Mackenbach J. The effects of ill health on entering and maintaining paid employment: Evidence in European countries. Journal of Epidemiology and Community Health. 2007;**61**:597-604

[55] European Agency for Safety and Health at Work (EU-OSHA). OSH in Figures: Young Workers—Facts and Figures. European Risk Observatory Report. Luxembourg; 2007

[56] Kemmlert K, Lundholm L. Slips, trips and falls in different work groups–with reference to age and from a preventive perspective. In: Applied Ergonomics. Apr 2001;**32**(2):149-153. PubMed PMID: 11277507

[57] EU-OSHA — European Agency for Safety and Health at Work, Gender issues in safety and health at work. A review. 2003a. Available from: http://osha.europa.eu/en/publications/reports/209 [Accessed: 06 March 2014]

[58] EU-OSHA — European Agency for Safety and Health at Work, New risks and trends in the safety and health of women at work. European Risk Observatory. A summary of an Agency report, European Agency for Safety and Health at Work. 2014. Available from: https://osha.europa.eu/en/publications/reports/

summary-new-risks-trends-oshwomen/view [Accessed: 06 March 2014]

[59] Verdonk P, Benschop YW, de Haes HC, Lagro-Janssen TL. From gender bias to gender awareness in medical education. Advances in Health Sciences Education: Theory and Practice. 2009;**14**(1):135-152. DOI: 10.1007/s10459-008-9100-z

[60] Kim AM, Tingen CM, Woodruff TK. Sex bias in trials and treatment must end. Nature. 2010;**465**:688-689

[61] Legato M. Principles of Gender-Specific Medicine. London: Elsevier Academic Press; 2010

[62] Bettio F, Verashchagina A. Gender Segregation in the Labour Market: Root Causes, Implications and Policy Responses in the EU, EC's Expert Group on Gender and Employment. Luxembourg: Publications Office of the European Union; 2009

[63] Cha Y, Thébaud S. Labor markets, breadwinning and beliefs: How economic context shapes men's gender ideology. Gender and Society. 2009;**23**(2):215-243

[64] Sorrentino E, Vona R, Monterosso D, Giammarioli AM. Gender issues on occupational safety and health. Annali dell'Istituto Superiore di Sanità. 2016;**52**(2):190-197. DOI: 10.4415/ANN_16_02_10

[65] Campos-Serna J, Ronda-Pérez E, Artazcoz L, Moen BE, Benavides FG. Gender inequalities in occupational health related to the unequal distribution of working and employment conditions: A systematic review. International Journal for Equity in Health. 2013;**12**:57. DOI: 10.1186/1475-9276-12-57

[66] Messing K, Stock SR, Tissot F. Should studies of risk factors for musculoskeletal disorders be stratified by gender? Lessons from the 1998 Québec health and social survey. Scandinavian Journal of Work, Environment & Health. 2009;**35**(2):96-112

[67] Messing K, Dumais L, Courville J, Seifert AM, Boucher M. Evaluation of exposure data from men and women with the same job title. Journal of Occupational Medicine. 1994;**36**(8):913-917

[68] Messing K, Silverstein BA. Gender and occupational health. Scandinavian Journal of Work, Environment & Health. 2009;**35**(2):81-83

[69] Food and Drug Administration. Guideline: General considerations for the clinical evaluation of drugs. US for the study and evaluation; 1977

[70] Arbuckle TE. Are there sex and gender differences in acute exposure to chemicals in the same setting? Environmental Research. 2006;**101**(2):195-204

[71] Franconi F, Brunelleschi S, Steardo L, Cuomo V. Gender differences in drug responses. Pharmacological Research. 2007;**55**:81-95

[72] Dimich-Ward H, Beking K, DyBuncio A, Chan-Yeung M, Du W, Karlen B, et al. Occupational exposure influences on gender differences in respiratory health. Lung. 2012;**190**(2):147-154. DOI: 10.1007/s00408-011-9344-x

[73] Carey MA, Card JW, Voltz JW, Arbes SJ Jr, Germolec DR, Korach KS, et al. It's all about sex: Gender, lung development and lung disease. Trends in Endocrinology and Metabolism. 2007;**18**(8):308-313

[74] Hoven H, Siegrist J. Work characteristics, socioeconomic position

and health: A systematic review of mediation and moderation effects in prospective studies. Occupational and Environmental Medicine. 2013;**70**(9):663-669

[75] Landsbergis PA, Grzywacz JG, LaMontagne AD. Work organization, job insecurity, and occupational health disparities. American Journal of Industrial Medicine. May 2014;**57**(5):495-515. DOI: 10.1002/ajim.22126. Epub 2012 Oct 16. Review. PubMed PMID: 23074099

[76] Peter R, Siegrist J, Hallqvist J, Reuterwall C, Theorell T. Psychosocial work environment and myocardial infarction: Improving risk estimation by combining two complementary job stress models in the SHEEP study. Journal of Epidemiology and Community Health. 2002;**56**(4):294-300

[77] d'Errico A, Cardano M, Landriscina T, Marinacci C, Pasian S, Petrelli A, et al. Workplace stress and prescription of antidepressant medications: A prospective study on a sample of Italian workers. International Archives of Occupational and Environmental Health. 2011;**84**(4): 413-424. DOI: 10.1007/s00420-010-0586-3

[78] ILO. Safetyand health at the heart of the future of work. 2019. Available from: https://www.ilo.org/wcmsp5/groups/public/---dgreports/dcomm/documents/publication/wcms_686645.pdf

Chapter 5

Linking Social Support with Job Satisfaction: The Role of Global Empowerment in the Workplace

Maria Helena Almeida, Alejandro Orgambídez Ramos and Carina Martinho Santos

Abstract

According to the concept of healthy organizations, three main interrelated components are considered: (1) structural resources for the execution of the task (e.g., autonomy) and social resources in the Working Group (e.g., social support); (2) healthy active professionals experiencing high levels of psychosocial well-being (job satisfaction); and (3) healthy organizational outcomes such as high performance and quality of service. The aim of this study is to examine the relationship between social resources (social support), structural resources (global empowerment) and job satisfaction. This study comprised a sample of 370 Portuguese healthcare professionals working in five stars private hospitals. A cross-sectional study was used. Data were collected based on personally administered surveys. An adjusted model of structural equations showed that job satisfaction was significantly predicted by social support and global empowerment. Additionally, employees' perception of empowerment can influence the relationship between social support and job satisfaction. Interventions based on support networks are decisive for increasing job satisfaction, but if the health institution offers structural conditions that foster global empowerment, this relationship is further strengthened. The cross-sectional design cannot highlight the causal relationships that longitudinal studies are more apt to do.

Keywords: social support, global empowerment, job satisfaction, caregivers

1. Introduction

Classically, organizational psychology has focused on studying negatively charged behaviors such as absenteeism, turnover and work stress, among others [1]. Today, a paradigm shift has allowed us to move from a negative perspective to a more positive one. As a result, positive psychology has emerged, as a movement that currently affects this research effort to better translate organizational projects, and expand and improve psychosocial well-being and quality of life in organizations [2]. This approach to positive psychology, at the level of enhancing the knowledge of a full organizational life, characterized by "positive employees" working in "positive organizations," is the ideal ground for characterizing so-called "healthy organizations." This concept of "healthy organizations" describes the great commitment made by organizations in defining strategies and actions characterized by good practices: These measures of a systematic, methodical, clearly delineated and proactive nature aim to improve the processes and outcomes of the organization so as to affect the welfare of both employees and the organization. This bravery on the part of the organization presupposes, however, the introduction of resources and the implementation of good experiences in the search for improvements in the work environment, in order to promote the health of employees and the financial health of the organization [3].

According to this model, healthy organizations consider the following three interrelated components: (1) social resources in the working group (e.g., social support) and structural resources for the execution of tasks (e.g., autonomy); (2) healthy active professionals experiencing high levels of psychosocial well-being; (3) healthy organizational outcomes such as high performance and quality of service [4]. In an attempt to explain each of these three dimensions: (1) strategies aimed at creating social resources in the working group are another strategy adopted by organizations, which can be provided in the form of social support. Social support can be provided to professionals by superiors [5, 6] or by the contributors themselves [7, 8]. Social support can, however, be more "emotional" in the more or less engaging configuration of active emotions with other members, both inside and outside the organization [9, 10]. But it can also have an "instrumental" configuration, through the more or less tangible manner of these interactions that facilitate the achievement of results, work execution, financial assistance and other facilitated aids or assets. In a second dimension, facilitated access to information channels (strategic, technical and practical), support (guidance, monitoring and feedback), resources (material, human or financial) and opportunities (to learn and grow in the organization) are organizational structures of which enable employees to perform their work activities and tasks freely, independently and autonomously. These structurally empowered environments, according to Kanter [11], are conducive to a global perception of greater autonomy and control by employees in carrying out their work, whose execution is more effectively predicted (global empowerment) [12]. Therefore, managers, by providing substantiated infrastructures (increased access to information, resources, support and opportunity), provide their subordinates with the support tools they need to perform their assigned activities and tasks with complete freedom and success. In turn, employees, as people who deserve this trust, must have the skills, abilities and talents necessary to perform these activities and tasks as successfully as possible. The end result is an overall perception of job effectiveness, based on the employees' perception of global empowerment. This assumption clearly requires managers to implement organizational strategies, which are essential for the promotion of truly empowered social support climates, to trigger high levels of well-being at work. (2) Active and healthy professionals with high levels of psychosocial well-being—employees' perception of global empowerment enables them to understand greater freedom and independence in the execution and management of their own work. It also allows them to feel more autonomous in decision-making without having to submit to higher authorization. This subjective state favors the appearance of job satisfaction. On the other hand, the social support given by superiors and peers results in professionals having a perception of being loved, cared for, esteemed and valued, based on a social network of mutual assistance [13]. The integration into social groups that establish friendship bonds and guarantee the necessary support to face the demands of work favors the appearance of positive attitudes at work, such as job satisfaction. These two dimensions, namely structured working conditions with sufficient resources to perform tasks independently (i.e., autonomy) and support working group (i.e., social support), are a reliable way to engage people in healthy activities. An increased sense of autonomy and a supportive climate allow employees to experience high levels of job satisfaction, psychosocial well-being and health at work [14, 15] in acting as risk-protecting agents against contracting diseases of a psychosomatic nature. (3) Healthy organizational outcomes, such as high performance and quality of service—the overall perception of empowerment and social support are decisive factors for more effective individual performance [10, 11], quality of life at work [16] and consequent achievement of personal goals [11], such as greater professional and/or personal achievement. In other words, job satisfaction is an indirect indicator of work efficiency and quality of service [12].

In the specific context of health, satisfaction is an important attitude that can benefit patient care [17, 18], particularly the quality of service provided [19–21]. Job satisfaction also has a positive effect on decreasing turnover intent [22, 23] and absenteeism [24, 25], results that, as we know, are detrimental to individuals and the organization. The above studies confirm and reinforce the general idea advocated by Kanter's structural empowerment theory (1977) that adequately trained work environments increase motivation and job satisfaction.

2. Relationship between job satisfaction and empowerment in carers

The impact of healthy work environments on job satisfaction attitudes has been evidenced in systematic literature reviews [26] and also in numerous studies [27–32].

These empirical efforts have been widely expressed in comparative country studies—mostly in North America, the United Kingdom and Western Europe—with samples of nurses, physicians and other health professionals e.g.,[17, 33]. Many efforts have been made to prove that professionals working in workplaces that are structured tend to exhibit high levels of job satisfaction (e.g., [34], updated in 2011; [35]). These studies have underlined that a person's perception of control and responsibility in a confrontation with work is an antecedent factor that determines the advent of job satisfaction. The findings also show a significant positive relationship between empowerment and job satisfaction (for a systematic review, see [36]), regardless of the design adopted or the sample described. Other studies have highlighted the importance of the correlation between empowerment and job satisfaction. This relationship is fundamental in promoting improvements in the quality of care provided but also in retaining people at the organization (e.g., [37–41]). All these studies reinforce the idea, originally defended by Kanter's [11] structural empowerment theory that empowered work environments foster motivation and job satisfaction.

3. Relationship between job satisfaction and social support in carers

In general, positive work attitudes (e.g., job satisfaction) establish a positive and meaningful relationship with work contexts characterized by active social dynamics [42, 43]. An illustrative example is social support, whether from superiors (e.g., [5, 6]) or from peers (e.g., [7, 44]). In particular, there are several systematic literature reviews (e.g., [34, 44–46]) that have shown that social support, whether provided by superiors or colleagues, is a predictive factor of job satisfaction, work involvement and carer commitment to the organization. These findings have shown that integration into social groups may not only enable the carer to establish bonds of friendship but also ensure the technical support he or she needs to meet the demands of the job. Thus, the positive social interactions that are established, not only between supervisors and health-care providers but also between the health-care providers and work colleagues (peers), in terms of orientation, follow-up, constructive feedback and focus on quality, can be a powerful source of job satisfaction.

This study intends to use only the first two allowances of the above model describing healthy organizations: (1) social resources in the working group (e.g., social support) and structural resources for the execution of tasks (e.g., autonomy); (2) healthy active professionals experiencing high levels of psychosocial well-being through job satisfaction. The permissive Healthy organizational outcomes such as high performance and quality of service would be a consequence of the attitude toward job satisfaction, which will not be evaluated in this study.

The goal is to understand the extent to which social resources in the work group (social support from superiors and peers) and the employees' perception of global empowerment correlate with job satisfaction (attitude, a characteristic of active and healthy professionals, who perceive a high psychosocial level).

4. Method

4.1 Design and sample

In a descriptive-correlational nature and following a quantitative methodology a model of structural equations was created to evaluate a sample composed of 370 health professionals—physicians, nurses, medical assistants and health technicians—from a private group five-star hospital health service in southern Portugal. It is a convenience sample that allows to draw valid conclusions, since it corresponds to about 50% of the universe of the target population.

With a mean age of 33.49 years (DP = 8.96), this sample is predominantly female (71.4%), in which the participants work in an inpatient regimen (40%), outpatient (38.4%) or another type of regimen (21.6%). The majority work full time (87.6%), in a shift work regime (78.9%) and in fixed schedule (21.1%). The majority of the participants (82.7%) worked in their profession and in the private health group for more than a year (75.7%), in an exclusive regime (80.8%), the rest (19.2%) work not only in this institution, but also in other institutions.

4.2 Procedure

The information was collected through a questionnaire survey. After the request for authorization, the ethics committees of the two hospitals of this private health group approved the study. The research questionnaires were then applied to health professionals who agreed to participate individually during normal working hours in a period of time created for this purpose. Each participant received the informed consent and the questionnaire in independent envelopes, in order to guarantee the desired anonymity and confidentiality at all moments of the information collection.

4.3 Instruments

Global empowerment-assessed through two items from the Global Empowerment subscale of Conditions of Work Effectiveness Questionnaire II (CWEQII2) by Laschinger et al. [12], on a Likert scale ranging from (1)—totally disagree and (5)—I totally agree.

Social support-evaluated through eight items of the social support subscale of the Job Content Questionnaire (JCQ) of Karasek and Theorell [47]: (a) social support of superiors (4 items); and (b) social support of peers (4 items) on a scale ranging from 1 (totally disagree) to 4 (totally agree).

Satisfaction in work-evaluated through the Job Satisfaction Scale (JSS) developed by Lima et al. [48] of eight items that ranged from 1 (totally disagree) to 7 (totally agree).

4.4 Data analysis

Descriptive statistics (mean, standard deviation, asymmetry and kurtosis), correlations between the variables under study (Pearson's coefficients), internal consistency coefficients (Cronbach's alpha) and the saturated structural equations model, tested to determine the relationships between global empowerment, social support of superiors and peers, and job satisfaction, were performed using the Software for Statistics and Data Science (STATA), version 13. To obtain a global representation of the relationship between social, superior and peer support, global empowerment and professional satisfaction, a saturated model of relationships was projected. This model was submitted to a structural equations test and redesigned from the standardized coefficients. The maximum likelihood (ML) method was used as a parameter estimation procedure to determine the effects (direct and indirect) and mediation [49, 50].

5. Results

5.1 Descriptive statistics and correlations

Table 1 presents the descriptive statistics (mean, standard deviation, asymmetry and kurtosis) and the correlations (Pearson's coefficient) of the studied variables, as well as the reliability coefficients and Cronbach's alpha of the scales used.

The mean value of the social support of the superiors was 3.41 (DP = 0.66) and the peers were 3.40 (DP = 0.56), indicating a tendentially positive level of support in the work environment. Global empowerment with an average of 3.42 (DP = 0.85) indicates a reasonable level of perceived global empowerment. Finally, health professionals are very satisfied at work (M = 4.53, DP = 1.02). The values of asymmetry and kurtosis are less than 1, not disrespecting the parameters that characterize normality in the data distribution ($|SK| < 3$ and $|KU| < 10$) [51]. The internal consistency

Variables	1	2	3	4
1. Social support from superiors	(0.92)			
2. Social support from peers	0.36**	(0.87)		
3. Global empowerment	0.52**	0.38**	(0.80)	
4. Job satisfaction	0.61**	0.47**	0.67**	(0.88)
Media	3.41	3.40	3.42	4.53
Standard deviation	0.66	0.56	0.85	1.02
Asymmetry	0.80	0.59	0.35	0.19
Kurtosis	0.35	−0.23	−0.03	−0.34

*Note: Alpha of Cronbach's values was presented in parentheses diagonally. All coefficients are significant "**" ($p < 0.01$).*

Table 1.
Descriptive statistics, correlations and reliability of the scales: social support of the superiors, social support of peers, Empowerment *global and job satisfaction (N = 370).*

of the scales used, assessed using Cronbach's alpha, show appropriate reliability [52]. As expected, a moderate, positive and very significant correlation was observed between work satisfaction and social support of supervisors ($r = 0.61$, $p < 0.01$), of colleagues ($r = 0.47$, $p < 0.01$) and global empowerment ($r = 0.67$, $p < 0.01$).

5.2 Mediation analysis

With the aim of presenting a global representation of the relationship between global empowerment, social support (of supervisors and peers), and professional satisfaction, the following relationship model was projected: (1) social support (superior and peers) were considered exogenous and predictive variables; (2) global empowerment, an endogenous and exogenous mediator variable; (3) professional satisfaction, endogenous variable and outcome. This model was empirically tested from an analysis of structural equations based on correlations. The analysis carried out had the following steps: (1) design of an over-identified model and (2) redesign of the model from the significant coefficients observed in the previous model, following the guidelines emitted by Acock [49]. For this purpose, a saturated structural equation model was tested, and items that did not present significant weights were then eliminated to determine the relationships between global empowerment, social support of superiors and peers, and satisfaction at work. The estimation of the effects (direct and indirect) as well as the mediation, used the maximum likelihood estimation (ML) method, the adjustment indices of the model and the Sobel test [49, 50].

Figure 1 shows a suitable final model. The adjustment index of the model, evaluated through the chi-square was significant (X^2 (2.1) = 79.271, $p < 0.01$). Values between 2 and 3 indicate a good fit of the model, so the values obtained showed an adjusted model.

The CFI and TLI indexes were all higher than 0.90 (CFI = 0.972; TLI = 0.953) as these values usually range from 0 to 1, the results show a satisfactory adjustment.

The adjustment indicator values, Standardized Root Mean Square Residual (SRMR) was less than 0.05 (SRMR = 0.035) and the coefficient Root Mean Square Error of Approximation (RMSEA) was from 0.087 [90% CI: 0.067–0.108], tend to hang between 0.05 and 0.08, being acceptable values up to 0.10, The results obtained are satisfactory [53, 54]. **Figure 1** also shows the standardized coefficients obtained in the structural equations model, as well as the explained variance (R^2) of the variables global empowerment and professional satisfaction. Global empowerment had a positive and significant predictive effect ($p < 0.01$) on the social support of the supervisor and the peers. Beta values were 0.48 for supervisory support and 0.24 for peer support. The total variance of global empowerment, explained

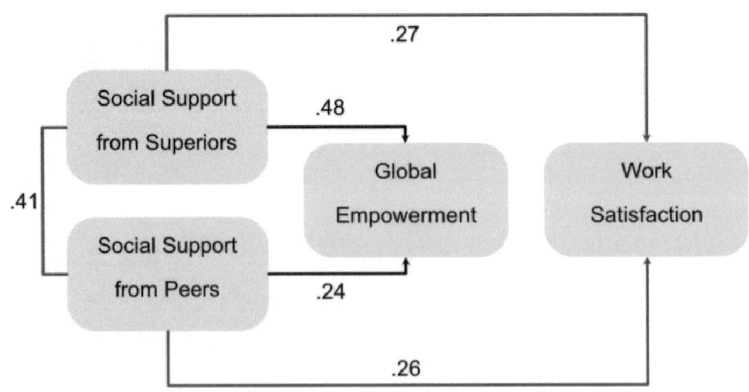

Figure 1.
Validated final model (N = 370). All coefficients are significant (p < 0.01).

Direct effects	**Coef.**	***SE***	**z**	**Beta**
Empowerment				
Support of the superiors →	0.62	0.07	8.49	0.48
Support of the peers →	0.36	0.09	3.95	0.24
Satisfaction				
Support of the superiors →	0.45	0.11	4.64	0.27
Support of the peers →	0.48	0.10	4.39	0.26
Empowerment →	0.87	0.11	8.17	0.53
Indirect effects				
Empowerment				
Support of the superiors →	(No path)			
Support of the peers →	(No path)			
Satisfaction				
Support of the superiors →	0.54	0.09	6.15	0.29
Support of the peers →	0.31	0.09	3.65	0.14
Empowerment	(No path)			
Total effects				
Empowerment				
Support of the superiors →	0.62	0.07	8.49	0.48
Support of the peers →	0.36	0.09	3.95	0.24
Satisfaction				
Support of the superiors →	1.01	0.10	10.29	0.56
Support of the peers →	0.79	0.10	6.56	0.40
Empowerment	0.87	0.11	8.17	0.53

Note: All coefficients are significant ($p < 0.01$).

Table 2.
Direct, indirect effect and total effect of the variables studied (N = 370).

by support from superiors and colleagues, was 38%. The professional satisfaction had a positive and significant predictive effect of global empowerment ($\beta = 0.53$, $p < 0.01$), support of supervisors ($\beta = 0.27$, $p < 0.01$) and support of peers ($\beta = 0.26$, $p < 0.01$). The total of the variance of professional satisfaction, explained by the global empowerment, support of superiors and peers, was 75%.

Regarding the mediating role, global empowerment mediated the influence of superior and peer support on job satisfaction. Support from superiors had a direct and indirect impact on professional satisfaction. Regarding the total effect of superior support on professional satisfaction, 48.2% (27/56) was direct, while 51.8% (29/56) was indirect. Peer support had a direct and indirect impact on job satisfaction. Concerning the total effect of peer support in professional satisfaction, 65% (26/40) was direct, while 35% (14/40) was indirect (**Table 2**).

6. Discussion

Assuming that job satisfaction is a fundamental attitude at work and an indirect indicator of efficiency and quality of service [12], it was intended to evaluate 370 health professionals working in a private hospital group in the south of Portugal. This study was supported by the Healthy and Resilient Organization Model (HERO; [3] in [4]), which provides an interrelationship between three main components: (1) resources in the working group (e.g., social support) and structural resources for the execution of tasks (e.g., autonomy); (2) healthy active professionals experiencing high levels of psychosocial well-being; and (3) healthy organizational results such as high performance and quality of service. In a similar way, it was intended to understand the extent to which social resources in the work group (social support from superiors and peers) and structural resources for the execution of tasks (globally empowered) relate to job satisfaction (experience high levels of psychosocial well-being) in a private hospital group in southern Portugal.

The results showed the role played by social relations in organizations through the positive and significant relationship between social support (from superiors and peers) and job satisfaction. This finding is corroborated by other studies [21, 44, 55, 56]. The predictive effect of social support (from superiors and peers) on job satisfaction was equally evidenced and similar to other findings in this area [34, 44, 46, 57]. This relevant role played by social relations, as has been highlighted in the literature [42, 43], has practical implications for superiors and colleagues, who play a key role in following up and giving constructive feedback to employees regarding the quality of care. Another interesting finding was the positive and significant relationship between global empowerment and job satisfaction. These results are consistent with Kanter's structural empowerment model [11] and with some investigations [12, 37–39, 41, 58, 59]. One practical implication of this result obtained through perceived global empowerment (highlighted by the perception of structured work environments characterized by providing easier access to information, resources, opportunities and support) is the need for managers to include this variable in the management of health institutions. This measure of creating healthy environments is crucial for promoting job satisfaction, which in turn is a critical factor in individual and organizational success. Another interesting finding was the mediating effect of global empowerment between social support (from superiors and colleagues) and job satisfaction. These findings are an indication that the global perception of effectiveness at work [12] is a factor that cannot be overlooked by these health institutions, since it can reduce the magnitude of the relationship between social support (independent variable) and job satisfaction (dependent variable). This shows the prevailing power of global empowerment from the point of view of working conditions.

The above shows that carers react emotionally to certain situations that arise from these structural conditions, which in turn influences their attitudes and behaviors [11, 12]. The findings also show the direct effect of social support on job satisfaction. Whether social support came from superiors or peers, the magnitude observed was very similar. This underlines the fact that social support fostered in the workplace, whether affective or instrumental in nature, has the genuine ability, by itself, to have an effect on job satisfaction without having to be mediated by other variables. An important implication for the managers of these institutions is that they should consider creating organizational environments that prioritize the integration of teams (whether superiors or peers) through fostering support and interaction. This strategy is decisive in promoting greater social support perceived in the institution in order to directly achieve job satisfaction. This measure, applicable to the participants in this study, can be extended to all health institutions. The creation of healthy social environments is essential in providing job satisfaction (indirect indicator of quality of service), in order to maximize available resources as well as the excellence of care provided. On the other hand, private health organizations, due to their exponential growth and the high demands from stakeholders, especially patients, who expect a timely response and quality of service, have the additional challenge of promoting employee well-being, so that they can feel motivated, supported and valued, and thus better meet the expectations and challenges that have been created for them.

The findings obtained in the present study should, however, be cautiously interpreted since the cross-sectional design does not allow conclusions to be drawn about the causality that a longitudinal study enables. The second limitation is that global empowerment is not the only mediating factor in the relationship between social support and job satisfaction, as there are other variables that will certainly play an equally relevant role, in mediating in this relationship. Studies of a longitudinal nature could help in better understanding the causal relationships between these variables in health care. A complementary qualitative analysis could also better explain the quality of the emotional and instrumental relationship between subordinates and peers.

7. Conclusion

The social support of superiors and peers and global empowerment seem to be two important determinants of job satisfaction in health care. The two types of social support, superior and peer, seem to affect job satisfaction both directly and indirectly through global empowerment. These findings are corroborated by Kanter's theory of structural empowerment [11]. The results show the relevance of social support (from supervisors and peers) that directly and indirectly influences positive attitudes such as job satisfaction. These findings suggest the need to invest in training and the development of social skills. These interventions are essential in fostering a culture of socio-affective support, follow-up and constructive feedback, to provide quality care but also to develop employee commitment to the organization. These results show the indispensability of an organizational culture characterized by greater effectiveness through the creation of infrastructures that enable the sharing of information, support, opportunities and resources that provide health professionals with greater autonomy and influence in their work and participation in decision-making, with a view to continuous improvement and professional development. A culture imbued with social support and empowerment fosters better management of the resources available at the unit and encourages motivation and job satisfaction by encouraging employees to feel needed, responsible and free to use their skills, abilities and skills. Moreover, it helps employees realize that they can count on organizational support, conveying the trust and respect that employees need to identify with the organization's goals and projects.

Author details

Maria Helena Almeida[1*], Alejandro Orgambídez Ramos[2] and Carina Martinho Santos[3]

1 School of Economics, University of Algarve, Faro, Portugal

2 Department of Social Psychology, Social Work, Social Anthropology and East Asian Studies, University of Malaga, Malaga, Spain

3 Private Hospital of Algarve (HPA), Faro, Portugal

*Address all correspondence to: halmeida@ualg.pt

References

[1] Salanova M. Organizaciones saludables y desarrollo de recursos humanos. Revista de Trabajo y Seguridad Social, CEF. 2008;**303**:179-214. Available from: www.want.uji.es›wp-content›uploads›

[2] Salanova M, Lorrens S, Martines IM. Aportaciones desde la psicología organizacional positiva para desarrollar organizaciones saludables y resilientes. Papeles del Psicólogo/Psychologist Papers. 2016;**37**(3):177-184. DOI: 10.1080/17439760.2015.1137628

[3] Salanova M, Llorens S, Cifre E, Martínez IM. We need a hero! Towards a validation of the Healthy & Resilient Organization (HERO) Model. Group & Organization Management. 2012;**37**:785-822. DOI: 10.1177/1059601112470405

[4] Salanova M, Martínez IM, Llorens S. Una mirada más "positiva" a la salud ocupacional desde la Psicología Organizacional Positiva en tiempos de crisis: Aportaciones desde el equipo de investigación WONT. Papeles del Psicólogo. 2014;**35**:22-30. DOI: 10.1037/t03624-000

[5] Bruce WM, Blackburn JW. Balancing Job Satisfaction and Performance: A Guide for Human Resource Professionals. Westport, Conn.: Quorum Books; 1992

[6] Vroom VH. Work and Motivation. Rev. ed. Malabar, FL: Robert E. Krieger Publishing Company; 1982

[7] Green J. Job satisfaction of community college chairpersons [Doctoral dissertation], Virginia Polytechnic Institute and State University; 2000. Electronic Theses & Dissertations Online, URN. Number etd-12072000-130914

[8] Maynard M. Measuring work and support network satisfaction. Journal of Employment Counselling. 1986;**23**: 9-19. DOI: 10.1002/j.2161-1920.1986.tb00187.x

[9] Aspinwall LG, Taylor SE. Modeling cognitive adaptation: A longitudinal investigation of the impact of individual differences and coping on college adjustment and performance. Journal of Personality and Social Psychology. 1992;**63**:989-1003. DOI: 10.1037/0022-3514.63.6.989

[10] Wanberg CR, Banas JT. Predictors and outcomes of openness to changes in a reorganizing workplace. Journal of Applied Psychology. 2000;**85**(1):132-142. DOI: 10.I037//0021-9010.85.1.132

[11] Kanter RM. Men and Women of the Corporation. New York, NY: BasicBooks; 1993

[12] Laschinger HK, Finegan J, Shamian J, Wilk P. Impact of structural and psychological empowerment on job strain in nursing work settings: Expanding Kanter's model. Journal Nursing Administration. 2001;**31**(5):260-272. DOI: 10.1097/00005110-200105000-00006

[13] Wills TA. Social support and interpersonal relationships. In: Clark MS, editor. Review of Personality and Social Psychology, Prosocial Behavior. Vol. 12. Thousand Oaks, CA, US: Sage Publications, Inc.; 1991. pp. 265-289

[14] Berkman LF, Syme L. Social networks, host resistance, and mortality: A nine-year follow-up study of alameda county residents. American Journal of Epidemiology. 1979;**109**(2):186-204. DOI: 10.1093/oxfordjournals.aje

[15] Pietrukowicz MCLC. Apoio social e religião: Uma forma de enfrentamento dos problemas de saúde

[Dissertação de Mestrado]. Fundação Oswaldo Cruz, Escola Nacional de Saúde Pública. 2001. Availabl from: http://portalteses.cict.fiocruz.br/transf.php?lng=pt&script=thes_thesislist&id=FIOCRUZENS

[16] Adams R. Empowerment, Participation and Social Work. 4ª ed. New York: Palgrave Macmillan; 2008

[17] Aiken LH, Clarke SP, Sloane DM, Sochalski J, Silber J. Hospital nurse staffing and patient mortality, nurse burnout, and job satisfaction. American Medical Association. 2002;**288**(16):1987-1993. DOI: 10.1001/jama.288.16.1987

[18] Wagner J, Cummings G, Smith DL, Olson J, Anderson L, Warren S. The relationship between structural empowerment and psychological empowerment for nurses: A systematic review. Journal of Nursing Management. 2010;**18**(4):448-456. DOI: 10.1371/journal.pone.0057570

[19] Armstrong KJ, Laschinger H. Structural empowerment, magnet hospital characteristics, and patient safety culture: Making the link. Journal of Nursing Care Quality. 2006;**21**(2):124-132. DOI: 10.1097/00001786-200604000-00007

[20] Gilbert S, Laschinger KH, Leiter M. The mediating effect of burnout on the relationship between structural empowerment and organizational citizenship behaviours. Journal of Nursing Management. 2010;**18**(3):339-348. DOI: 10.1111/j.1365-2834.2010.01074.x

[21] Orgambídez-Ramos A, Borrego-Alés Y, Vázquez-Aguado O, March-Amegual J. Structural empowerment and burnout among Portuguese nursing staff: An explicative model. Journal Nursing Management. 2017;**25**(8):616-623. DOI: 10.1111/jonm.12499

[22] Simon M, Müller BH, Hasselhorn HM. Leaving the organization or the profession—A multilevel analysis of nurses' intentions. Journal of Advanced Nursing. 2010;**66**(3):616-626. DOI: 10.1111/j.1365-2648.2009.05204.x

[23] Tsai Y, Wu SW. The relationships between organisational citizenship behaviour, job satisfaction and turnover intention. Journal of Clinical Nursing. 2010;**19**:3564-3574. DOI: 10.1111/j.1464- 0597.2009.00414.x

[24] Abiodun AJ, Osibanjo AO, Adeniji AA, Iyere-Okojie E. Modelling the relationship between job demands, work attitudes and performance among nurses in a transition economy. International Journal of Health Management. 2014;**7**(4):257-264. DOI: 10.1002/dc.20361

[25] Siu OL. Predictors of job satisfaction and absenteeism in two samples of Hong Kong nurses. Journal of Advanced Nursing. 2002;**40**(2):218-229. DOI: 10.1046/j.1365-2648.2002.02364.x

[26] Wei H, Sewell KA, Woody G, Rose MA. The state of the science of nurse work environments in the United States: A systematic review. International Journal of Nursing Sciences. Open Access funded by Chinese Nursing Association. 2018;**5**(3):287-300

[27] Baernholdt M, Mark BA. The nurse work environment, job satisfaction and turnover rates in rural and urban nursing units. Journal of Nursing Management. 2009;**17**(8):994-1001. DOI: 10.1111/j.1365-2834.2009.01027

[28] Friese CR. Nurse practice environments and outcomes: Implications for oncology nursing. Oncology Nursing Forum. 2005;**32**(4):765-772. DOI: 10.1111/j.1475-6773.2007.00825.x.1145

[29] Kotzer AM, Koepping DM, LeDuc K. Perceived nursing work environment of acute care pediatric nurses. Pediatric Nursing. 2006;**32**(4):327-332. DOI: 10.12968/bjon.2018.27.4.197

[30] Laschinger HKS, Nosko A, Wilk P, Finegan J. Effects of unit empowerment and perceived support for professional nursing practice on unit effectiveness and individual nurse well-being: A time-lagged study. International Journal of Nursing Studies. 2014;**51**(12):1615-1623. DOI: 10.1016/j.ijnurstu.2014.04.010

[31] Teclaw RKO. Nurse perceptions of workplace environment: Differences across shifts. Journal of Nursing Management. 2015;**23**(8):1137-1146. DOI: 10.1111/jonm.12270

[32] Ulrich S. Applicants to a nurse-midwifery education program disclose factors that influence their career choice. Journal of Midwifery & Women's Health. 2009;**54**(2):127-132. DOI: 10.1016/j.jmwh.2008.12.015

[33] Leiter P, Maslach C. Nurse turnover: The mediating role of burnout. Journal of Nursing Management. 2009;**17**:331-339. DOI: 10.1111/j.1365-2834.2009.01004.x

[34] Lu H, While AE, Barriball KL. A model of job satisfaction of nurses: A reflection of nurses' working lives in mainland China. Journal of Advanced Nursing. 2007;**58**:468-479. DOI: 10.1111/j.1365-2648.2007.04233.x

[35] Carvalho CMS, Gouveia AL, Pinto CAB, Mónico LSM, Correia MMF, Parreira PMSD. Empowerment on healthcare professionals: A literature review. Psychologica. 2017;**60**(2):45-63. DOI: 10.14195/1647-8606-60-2-3

[36] Squires H, Chilcott J, Akehurst R, Burr J, Kelly M. A framework for developing the structure of public health economic models. International Journal of Public Health. 2016;**19**(5):588-601. DOI: 10.1016/j.jval.2016.02.011

[37] Laschinger HKS, Almost J, Tuer-Hodes D. Workplace empowerment and magnet hospital characteristics: Making the link. Journal Nursing Administration. 2003;**33**(7-8):410-422. DOI: 10.1097/00005110-200307000-00011

[38] Laschinger HKS, Leiter M, Day A, Gilin D. Workplace empowerment, incivility, and burnout: Impact on staff nurse recruitment and retention outcomes. Journal of Nursing Management. 2009;**17**(3):3302-3311. DOI: 10.1163/9789004281196-005

[39] Laschinger HKS, Leiter M, Day A, Gilin-Oore D, Mackinnon SP. Building empowering work environments that foster civility and organizational trust. Nursing Research. 2012;**61**(5):316-325. DOI: 10.1097/NNR.0b013e318265a58d

[40] Lee G, Kim PB, Perdue RR. A longitudinal analysis of an accelerating effect of empowerment on job satisfaction: Customer-contact vs. non-customer-contact workers. International Journal of Hospitality Management. 2016;**57**(1):1-8. DOI: 10.1108/MD-02-2016-0089

[41] Sun N, He Z, Wang LB, Li QJ. The impact of nurse empowerment on job satisfaction. Journal of Advanced Nursing. 2009;**65**(12):2642-2648. DOI: 10.1111/j.1365-2648.2009.05133.x

[42] Jayasuriya R, Whittaker M, Halim G, Matineau T. Rural health workers and their work environment: The role of interpersonal factors on job satisfaction of nurses in rural Papua New Guinea. BMC Health Services Research. 2012;**12**:-156. DOI: 10.1186/1472-6963-12-156

[43] Judge TA, Kammeyer-Mueller JD. Job attitudes. Annual Review of

Psychology. 2012;**63**:341-367. DOI: 10.1146/annurev-psych-120710-100511

[44] Chiaburu DS, Harrison DA. Do peers make the place? Conceptual synthesis and meta-analysis of coworker effects on perceptions, attitudes, OCBs, and performance. Journal of Applied Psychology. 2008;**93**:1082-1103. DOI: 10.1037/0021-9010.93.5.1082

[45] Lambert EG, Minor KI, Wells JB, Hogan NL. Social support's relationship to correctional staff job stress, job involvement, job satisfaction, and organizational commitment. The Social Science Journal. 2016;**53**(1):22-32. DOI: 10.1016/j.soscij.2015.10.001

[46] Utriainen K, Kyngas H. Hospital nurses' job satisfaction: A literature review. Journal of Nursing Management. 2009;**17**(8):1002-1010. DOI: 10.1080/09243450902883920

[47] Karasek R, Theorell T. Healthy Work. Stress, Productivity, and the Reconstruction of Working Life. New York, NY: BasicBooks; 1990

[48] Lima ML, Vala J, Monteiro MB. Culturas organizacionais. In: Vala MB, Monteiro MB, Lima ML, Caetano A, editors. Psicologia Social e das Organizações—Estudos em Empresas Portuguesas. Lisboa: Celta Editora; 1994

[49] Acock AC. Discovering Structural Equation Modeling Using Stata. Texas, US: StataCorp LP; 2013

[50] Hayes AF. Introduction to Mediation, Moderation, and Conditional Process Analysis: A Regression-Based Approach. New York, NY: The Guilford; 2013

[51] Marôco J. Análise de Equações Estruturais. Perô Pinheiro: Report Number; 2010

[52] Nunnally JC. Psychometric Theory. 2nd ed. New York, NY: McGraw-Hill; 1978

[53] Hu LT, Bentler PM. Cutoff criteria for fit indexes in covariance structure analysis: Conventional criteria versus new alternatives. Structural Equation Modeling. 1999;**6**:1-55. DOI: 10.1080/10705519909540118

[54] Ullman JB. Structural equation modeling. In: Tabachnick BG, Fidell LS, editors. Using Multivariate Statistics. 5th ed. Boston: Pearson Education; 2007

[55] Pohl S, Galletta M. The role of supervisor emotional support on individual job satisfaction: A multilevel analysis. Applied Nursing Research. 2017;**33**:61-66. DOI: 10.1016/j.apnr.2016.10.004

[56] Yuh J, Choi S. Sources of social support, job satisfaction, and quality of life among childcare teachers. The social Science Journal. 2017, 2017;**54**:450-457. DOI: 10.1016/j.soscij.2017.08.002

[57] Owen DC, Boswell C, Opton L, Franco L, Meriwether C. Engagement, empowerment, and job satisfaction before implementing an academic model of shared governance. Applied Nursing Research. 2018;**41**(February):29-35. DOI: 10.1016/j.apnr.2018.02.001

[58] Almeida MH, Orgambídez-Ramos A, Batista P. Workplace empowerment and job satisfaction in portuguese nursing staff: An exploratory study. Central European Journal of Nursing and Midwifery. 2017;**8**(4):749-755. DOI: 10.15452/CEJNM.2017.08.0028

[59] Dahinten VS, Lee SE. Disentangling the relationships between staff nurse's workplace empowerment and job satisfaction. Journal of Nursing Management. 2016;**24**(8):1060-1070. DOI: 10.1111/jonm.12407

Chapter 6

Hand Hygiene Practices in Public Restrooms: Effects and Proposed Solutions

Bankole K. Fasanya, Moruf Adegbite, Maged Mikhail and George L. Stefanek

Abstract

Human safety is a popular ongoing research area in personal hygiene. Researchers are mostly apprehensive about how to protect humans from different hazards in the environment. Thus, guidelines developed for good hand wash practices in the public restrooms have showed little or no impact on human behavior. This research examined hand wash acts in the public restrooms and proposed possible solutions to improve the practice. There are 427 people who participated in the study. Participant age ranged from 18 years old and upward: statistically, female, 63%; male, 35%; and unidentified, 2%. Descriptive statistics revealed 99.5% respondents approved restroom redesign for appropriate hand hygiene practice, while 49% suggested restroom device automation. Inferential statistics results on redesign with a Welsh t-test were statistically significant ($t=1.967$, $df=300$, $p<0.0001$; $t=1.990$, $df=80$, $p<0.0001$; $t=1.9746$, $df=163$, $p<0.0001$). Findings showed that hand hygiene guidelines and recommendations are insufficient to ensure proper promotion of hand wash practices in restrooms. This study concluded that good hand wash practices in public restrooms could be ergonomically redesigned to include a visual and auditory alert that reminds users to wash their hands after restroom usage and to include hand wash practice in school curriculum. The findings from this study could be applicable in restaurants, schools, and bars, to manage and control transmission of disease through direct hands contacted with infectious diseases in the restrooms.

Keywords: hygiene, human health, restroom, handwashing, ergonomics

1. Introduction

Human hygiene is considered as one of the most effective ways of preventing diseases, and germs. Center for diseases control and prevention (CDC) associated many diseases and sicknesses to poor body hygiene [1, 2]. One key body part that requires constant and continuous hygiene to prevent human infection from germs and diseases is hands. In 2004, one of the US FDA studies revealed that food establishments were frequently out of compliance with the food code requirements for proper and adequate handwashing. In the study, the percentage of food

IntechOpen

establishments observed to be out of compliance with handwashing requirements ranged from 34% in hospitals to 73% in full-service establishments [3]. Our world today faces alarming rates of diseases related to hygiene and the quick spread of germs via contact of improper hand hygiene. According to the CDC [4], it is established that keeping hands clean is one of the most important steps to avoid sickness and spreading of germs and diseases to others. Hand hygiene is the most effective measure for interrupting the transmission of microorganisms, which cause infection both in the community and in the healthcare setting. Therefore, using epidemiology model, transmission of diseases and germs through poor hand hygiene practices could be reduced in the society. The use of restrooms and other everyday activities regularly expose a human to germs; therefore, the need to explore other factors surrounding handwashing for hygiene purposes is essential.

The knowledge of handwashing as a measure of hand and personal hygiene is not new, as it has been successfully transferred from generations to generations. Handwashing has been linked to culture and religion, see Staub [5]. The United States of America officially recommended that healthcare workers (HCWs) should wash their hands with soap for 1–2 min before and after patient contact, see Coppage [6]. In 1975, CDC released handwashing guidelines and practice in the healthcare (hospitals) and later modified the original version in 1985 [7, 8]. Research has revealed that the revised CDC handwashing guideline has been expanded for better practice in all organizations [9–11].

While we can say training people on the handwashing routine is important and it has helped improving hand hygiene practice, the attention to hand hygiene in health and safety should go beyond educating people commonly known in the healthcare industry. According to Jang et al. [12], healthcare workers' workload, other job interruptions, and overly conservative guidelines make it difficult to adhere to hand hygiene. The same authors concluded that it is imperative to study other factors on proper handwashing practice to make handwashing hygiene a lifestyle other than a routine. The process of hand hygiene could be regarded as cumbersome and demanding for many people due to workload and other conditions like availability of toiletries, environmental and structural conditions. It is no doubt that many people know that hand hygiene what is the most effective measure to prevent microbial pathogen cross-transmission and other healthcare-associated infections but wasn't enough to get people to do it and practice the routine that leads to good handwashing practice. World Health Organization (WHO) reported that several continuous evaluations and good team methodologies have helped compliance but are yet to sustain in specific critical communities and healthcare areas [13]. The study further shows that mentoring aside other handwashing routines could be an excellent way to make hand hygiene a lifestyle.

Experts have proven that hand hygiene is the most effective measure for interrupting the transmission of microorganisms, which cause infection both in the community and in the healthcare setting. Using hand hygiene as a training measure of reducing disease is unlikely to be successful when other factors in infection control, such as environmental hygiene, crowding, staffing levels, and education, are inadequate. Therefore, the way people use restrooms and regularly exposed to germs in everyday activities makes it essential to explore other factors surrounding handwashing that could encourage hand and body hygiene in general.

The lack of appropriate infrastructure, religious beliefs, and workplace conditions were the major influencers on the low compliance to handwashing hygiene among the health professionals [14–16]. It is about time to start focusing on the influencers on low complaint rates.

1.1 Children hand hygiene

About 1.8 million children under the age of 5 die due to diarrhea and pneumonia every year [17]. Diarrhea and pneumonia are the top two killers among children all over the world [17]. The same study showed how handwashing is the most effective way to prevent people in the world from dying through diseases and much other life-threatening conditions from hand contamination. Handwashing with soap has a strong capability to protect and shield about one out of every three young children who get sick with diarrhea [18, 19], and practically one out of five young children with respiratory infections like pneumonia [19]. Another research shows handwashing in educational institutions and access to water and soap in schools could improve student health [20, 21]. Children's exposure to proper handwashing from early life might help to improve their development in some settings [22]. If hand washing could affect children around the world this much, then, proper handwashing should be included in all organization culture.

If hand hygiene can be included in daily human behavior, it automatically becomes a lifestyle and not just a routine, because humans can tire of routine with time, but a lifestyle is part of human behavior that cannot be easily broken. In 2005, Jumaa highlighted areas needed for further research on proper handwashing, which include environmental conditions, people acts etc. and further concluded that cultural and behavioral issues also contributed to the poor practice of hand hygiene. This study investigated other factors influencing public's poor handwashing habits, suggested better restroom designs, and provided recommendations for improvement.

Objectives of the study

- Develop a questionnaire that captures public opinion on the use of public restrooms or school restrooms.
- Propose a better design structure based on survey feedback and design of prototype.
- Evaluate prototype from user perception.

2. Working methods

This study was conducted in a university environment. The university is in Hammond and Westville, Northwest Indiana US. Data were collected through survey. The survey was a self-developed questionnaire adapted from a standardized resource online. Survey was reviewed by three experts, two health practitioners, and one safety specialist to validate the contents for the purpose of the research. The questions were presented in the form of Likert scale options. The IRB office of the university where the study was conducted approved the protocol before survey distribution. The survey was distributed via email to all students, faculty, and all the university employees. Four hundred and twenty-seven (427) participants, including 246 students, 109 university employees, and 72 faculty, participated in the study. Participants include 147 males, 270 females, and 9 people who preferred not to indicate their gender. All the participants took the same survey with the same preferences given to all. The research was divided into three parts. The first part is the questionnaire, research on appropriate solutions and redesign for appropriateness. The survey questions focused on the toilet's settings, design, user habit, and toiletries availability.

3. Data analysis and results

3.1 Data analysis

Data compilation was done with the use of Excel® version 365 ProPlus. The use, cleanliness, structure, and factors that affect handwashing practices were evaluated using simple descriptive statistics and inferential statistics (Welch T-tests). Descriptive statistics was used to assess the differences in the proportions of participants reporting specific handwashing practices by gender. At the beginning of the analysis, the data were checked for normality. As expected, the percentage who had used the college restrooms one time or more was found to be 100% (out of which 35% were male, 64% female, and 2% preferred not to say). **Table 1** details the frequency distribution of the participants in percentage. Analysis shows that the percentage of those who used the restrooms always was the highest with 71% compared with those who sometimes or rarely used the restroom, that is, 14%.

The percentage of those who are comfortable using the school restroom was 44% based on different levels of comfortability while the percentage of those who are rarely or not comfortable is 34%. About 88% of the participants reported to always wash their hands after the use of restrooms. Approximately 99% agreed to have seen someone at some point walking out of the restroom without washing their hands, 85% strongly agreed that hand and body hygiene is paramount after the use of public and private restrooms. Furthermore, 72% strongly agreed that proper handwashing will prevent one from many diseases and sicknesses, 22% agreed to some degree while only 3% somewhat disagreed. In another question, participants were asked if they have received any form of hand hygiene training in the last 3 years. The responses show that 50% had received different forms of training in the last 3 years. The use of alcohol-based hand rub for hand hygiene also shows about 67% at different level of agreement.

Only 215 participants responded to the open-ended question, the responses showed 48.8% strongly suggested restroom redesign to improve hand hygiene and reduce germs transmission. Approximately 14% believe that automated hand-sanitizer machines should be installed in the restrooms. Further, only 3% of the participants responded that the inclusion of handwashing training as part of the school curriculum and regular cleaning of the university community restrooms would improve hygiene practices.

	Level	Responses
Gender	Male	35%
	Female	63%
	Prefer not to say	2%
Category	Student	58%
	Faculty	17%
	University employee	25%
Age	18–25	40%
	25–30	8%
	30+	52%

Table 1.
Participants percentage distribution.

3.2 Results

Survey questions were divided into two categories, the first part focused on individual hand hygiene and the second part focused on restroom restructure or redesign for health safety purposes of the community. **Figure 1** is the graphical representation that shows the significances of some major factors that could potentially affect college restrooms and handwashing hygiene among college students:

The purpose of this study is to investigate factors influencing poor handwashing practices in the community, especially among the younger generation. These days, poor handwashing practices and inadequate body hygiene have been attributed to different illnesses around the world and have increasingly expanded the spreading of the present pandemic case called COVID-19. Therefore, the results of this study could be instrumental to promote good handwashing practices. Promotion of good handwashing will prevent viral infection and reduce exponential chances of spreading any illness outbreak. Handwashing hygiene should be taken beyond restrooms, but also to prevent viral infections, especially those that spread through droplets from coughs and sneezes. In this situation, proper handwashing is the first line of measure.

Four survey questions that specifically addressed restroom redesign and restructure were separately analyzed using descriptive and inferential statistics (Welch T-test). Participants' responses on the question "Do you think public restrooms should be structured and well designed for Heath safety?" were analyzed. The participants that neither agreed nor disagreed (in-between) were eliminated in order to define the significance of those that agreed or disagreed. The number of people in this category was found to be 10, constituting 2% of the total participants. Group 1 was considered people who disagreed and group 2 as people who agreed with restroom redesign.

Table 2 shows that only two participants disagreed to the opinion that public restrooms on campus should be restructured and well redesigned for health safety

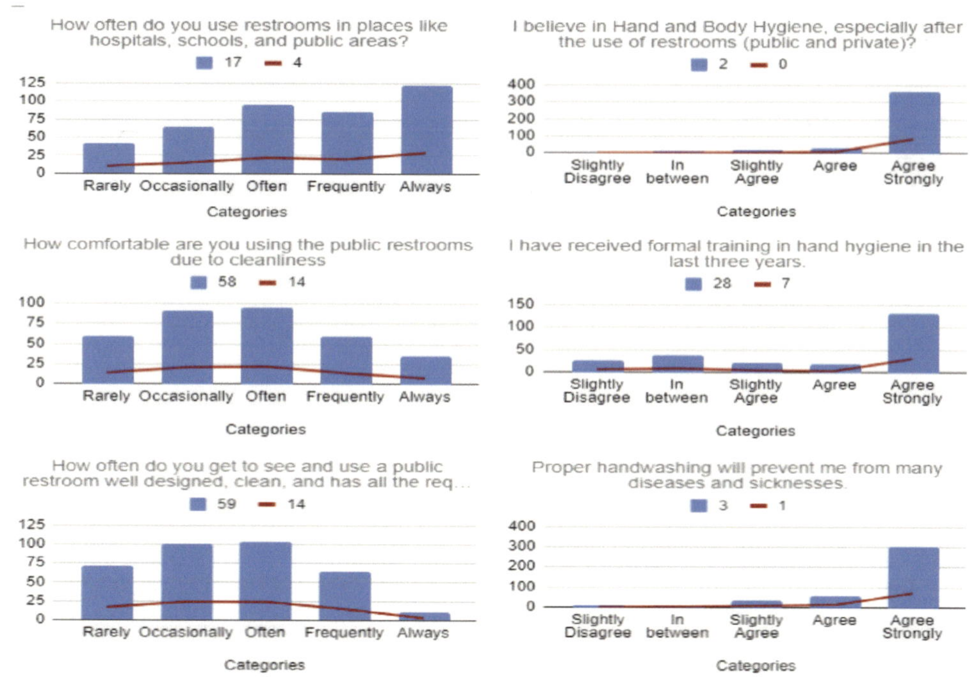

Figure 1.
Graphical representations of survey responses.

of the users, while 416 agreed to the opinion that university community restrooms should be restructured and redesigned for health safety (**Table 3**).

Based on the descriptive analysis of the user's opinion, it is concluded that restrooms should be redesigned for health safety. **Figure 2** reveals the relationship between the participants that agreed and disagreed that restrooms' redesign would improve the health safety of the users.

Questions 3, 6, and 12 were analyzed using Welch t-test to further determine the significance of redesigning and restructuring university community restrooms for health safety. As shown in **Table 4**, participants' response to question 3 reveals ($t = 1.967903$, $df = 300$; $p < 0.0001$) to question 6 ($t = 1.99006$, $df = 80$; $p < 0.0001$) and to question 12 ($t = 1.97462$, $df = 163$; $p < 0.0001$).

Categories	Participants' answer
Disagree	2
Neither nor	9
Agree	416
Total	427

Table 2.
Participants response to question 4.

Groups	Categories	Response	Percentage %
G1	Disagreed	2	0.5
G2	Agreed	416	99.5
Total		418	

G1 = Group 1; G2 = Group 2.

Table 3.
Participants' response to question 4 in percentages.

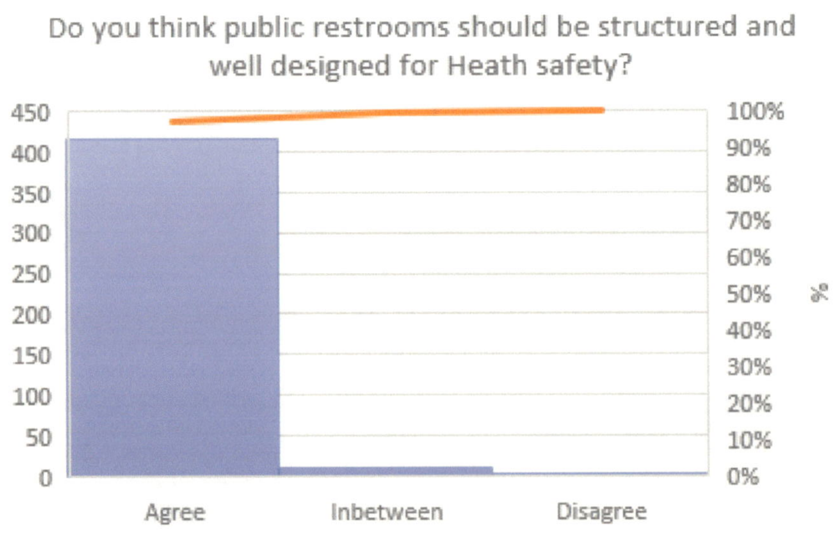

Figure 2.
Graphical representation of participants' response to question 4.

Questions	Method	Variance	df	t-value	P(T<=t)
3	Welch	Unequal	300	1.967903	< 0.0001
6	Welch	Unequal	80	1.99006	< 0.0001
12	Welch	Unequal	163	1.97462	< 0.0001

Table 4.
Summary of the Welch unpaired T-test.

4. Findings

1. From the participant's opinions, visual and auditory alerts will be necessary for most of the public restrooms to enhance user's awareness of handwashing after use.

2. Redesign of most restrooms was another issue raised by the respondents if proper hygiene levels must be reached for safety purposes.

3. Automated handwashing devices like an automated sink with soap dispenser, auto sensor water, and hand sanitizer pump dispenser were also recommended to be made available.

4. A redesign of the restroom doors to be touchless (auto open and close) without touch or ergonomical design (i.e., pull to go in and push to exit) as the participants felt that touching might increase the chances of contacting germs.

5. A high percentage of the participants also believe that the current hand dryer is unhealthy, breeds germs, and is against proper hygiene and health safety.

6. Respondents also mention the inclusion of hygiene training into college's education curriculums in all levels of education.

Ergonomically redesign doors/restroom system was suggested in order to fit the operating process of the restroom to the users' capability. This could be perfect as stated in Fasanya and Shofoluwe's [23] finding that fitting job to worker capability improved worker performances.

4.1 Redesigning option

After a careful review of the findings, it is decided that the inclusion of auditory and visual alert sensor in the **urinal bowl** area and **toilet compartment** will improve the handwashing practice.

4.1.1 Operation process of sensor

The visual and auditory alerting signals will assist in reminding restroom users to wash their hands after use. Visual and auditory alerts have been known for being vital reminders of activity in different areas of life such as in traffic control (transportation industry). Visual and alert methods have also played important roles in many other situations relating to alerting and reminding users. It has been used in the airplane for almost everything possible, especially in alerting, reminding, and giving the passengers directions when needed. According Papastavrou and Lehto

[24], visual and auditory alerts help in the detection of anticipated stimuli. This study suggested for the design shown in **Figure 3** to ensure proper reliabilities and the alerts required for the safe use of the restrooms. The design is a computer-based script that senses restroom flush and gently nudges and reminds the users to wash their hands. The process involves the following.

4.1.2 Design processes

It involves designing a computer screen algorithm that will help to auto-sense sides. The machine design can be achieved by using C# combined with light and sound using sensor fusion with computer vision to sense flushing sound and press of restroom flushes. **Figure 3** also shows the sample of visual message that would be displayed on the screen.

This device is an auditory and visual alert device coded with C# and with a motion and sound sensor to sense either the motion or flush sound before displaying a message as shown in **Figure 3**.

The visual and auditory alert system design would be placed in two locations as shown in the **Figure 4**—the device with the "Message Area" caption here is a LED Message display board designed and computer coded for auditory and visual alert for the users at the sound of restroom flush. The device should be a computer coded

Figure 3.
Message display.

Figure 4.
Restroom message display design.

with the **passive infrared (PIR) sensor** to sense and detect body heat *(***infrared energy**) and the most widely used motion sensor, at the sound of the flush. Pleasant messages are to be included in the alert to remind the users to wash their hands: messages like "Do you know handwashing hygiene increases health safety? Don't forget to wash your hands? Washing Hands prevents from deadly diseases, etc." The application that controls the device is suggested to be designed with C# and designed with LED message display board and passive infrared sensor. It is important for it to be both visual and auditory to accommodate visually or aurally challenged individuals.

4.2 Post analysis results

Findings from this study had led the university management to include in all restrooms hand soap dispensers and gradually work on how to restructure the rooms to be ergonomically fit for the users. From the management comments, less attention has been given to restroom designs of all things in the academic environment. Meanwhile, the outcome of this study has proved that less important things in management perspective might be a huge factor to promote health and safety of both students and the employees.

5. Discussions and conclusion

5.1 Discussions

Different studies have concluded that there are other factors affecting hand hygiene behavior. For example, Zimakoff et al. [25] concluded by identifying few other factors affecting hand hygiene such as skin irritation and dryness as the leading factors affecting handwashing in health care. The same authors affirmed that there are other possible factors not covered in the scope of the research. Likewise, in 1982, Larson and Killien concluded that it is imperative to identify factors that are the antecedents to whether the individual decides to wash one's hands or not as they are critical in the prevention or intervention plans to improve handwashing practice. The same authors further ascertained that most emphases are placed on the importance of handwashing instead of other factors affecting people's behavior toward handwashing compliance. The results of this study supported that there are other factors different from training people and showing people how to practice hygiene properly. Researchers are now focusing on the growing literature about the other factors affecting hand hygiene behavior.

Day by day, researchers are focusing on other factors that could affect hand hygiene behavior, none or few have looked into the structure and redesign of restrooms. This study investigated the people's experience in the restroom, why handwashing compliance is low, and other factors, which could affect hand hygiene behavior. A significant proportion of the participants requested for restroom redesign for their health safety. A significant finding from this study is that about 83% of the participants suggested alert systems to remind restroom users to wash hands after use. The participants also reported the significance of other factors like an automated sink, soap dispenser, water dispenser, dryer devices as well as ergonomically designed doors. The participants unanimously believe that ergonomically designed door (pull and push type) would reduce touch and greatly enhance health safety. The above data results reflect restroom equipment, structure, cleanliness, and comfortability have significant effects on hand hygiene behavior. The results from this study supported other researchers who found that there are other factors affecting hand hygiene behavior, rather than just training and those factors should be focused on in order to improve handwashing practices.

5.2 Conclusion

The handwashing procedure is a little monotonous, especially for healthcare professionals, and can be frustrating or become things of no interest when relevant factors are not available and very challenging in different ramifications. The low compliance of handwashing is a significant issue around the world as reported in several literatures. Appropriate handwashing practices can reduce the risk of foodborne illness and prevent transmission of viral infections, especially those that spread through droplets from coughs and sneezes. Besides, among many other hygiene practices, poor handwashing is the most common practice among the younger generation. The message and the information about handwashing or hand hygiene has been around for over 200 years, yet the level of compliance is low as revealed from this study. Thus, this study details another factor affecting handwashing hygiene after the use of public restrooms and suggests how hand hygiene could be improved in human daily behavior. The findings of this study revealed that restroom designs and structures have significant effects on hand hygiene behavior based on participant's opinions from the survey. Redesign to include visual and auditory alerts will be necessary for most of the public restrooms to enhance user's awareness for handwashing after use. Findings from this study suggest for a restroom redesign to include auto-sensor soap and water dispenser, dryers, and touchless doors. Findings further revealed that inclusion of hygiene training in all college's education curriculums at every level would encourage user's handwashing hygiene lifestyle and behavior. The findings from this study could help identify the design structures for ensuring more compliance with handwashing hygiene and health safety practice among the public. Further research is needed in this direction to investigate how other factors aside from the aforementioned could potentially discourage individuals from making hand hygiene a daily behavior.

Author details

Bankole K. Fasanya*, Moruf Adegbite, Maged Mikhail and George L. Stefanek
College of Technology, Purdue University Northwest, Hammond, Indiana, USA

*Address all correspondence to: fbankole@pnw.edu

References

[1] Show Me the Science—How to Wash Your Hands. Handwashing. CDC. 2016. Available from: https://www.cdc.gov/handwashing/show-me-the-science-handwashing.html [Retrieved: 08 September 2019]

[2] Water, Sanitation & Environmentally-Related Hygiene. Hygiene. Healthy Water. CDC. 2016. Available from: https://www.cdc.gov/healthywater/hygiene/index.html [Retrieved: 13 September 2019]

[3] Food and Drug Administration. FDA Report on the Occurrence of Foodborne Illness Risk Factors in Selected Institutional Foodservice, Restaurant, and Retail Food Store Facility Types. 2004. Retrieved from: https://wayback.archiveit.org/7993/20170406023011/https://www.fda.gov/downloads/Food/GuidanceRegulation/RetailFoodProtection/FoodborneIllnessRiskFactorReduction/UCM423850.pdf

[4] Show Me the Science—Why Wash Your Hands? Handwashing. CDC. 2018. Available from: https://www.cdc.gov/handwashing/why-handwashing.html [Retrieved: 05 September 2019]

[5] Staub E. The Roots of Evil: The Psychological and Cultural Origins of Genocide and Other Forms of Group Violence. United Kingdom: Cambridge University Press; 1989

[6] Coppage CM. Hand Washing in Patient Care [Motion Picture]. Washington, DC: US Public Health Service; 1961

[7] Steere AC, Mallison GF. Handwashing practices for the prevention of nosocomial infections. Annals of Internal Medicine. 1975;**83**(5):683-690

[8] Fevero MS. CDC guideline for handwashing and hospital environmental control, 1985. Infection Control. 1986;**7**:231-243

[9] Hospital Infection Control Practices Advisory Committee. Recommendations for preventing the spread of vancomycin resistance. Infection Control and Hospital Epidemiology. 1995;**16**(2):105-113

[10] Garner J. Guideline for isolation precautions in hospitals. Infection Control and Hospital Epidemiology. 1996;**17**(1):54-80. DOI: 10.1017/s0195941700006123

[11] Boyce JM, Pittet D. Guideline for hand hygiene in health-care settings: Recommendations of the healthcare infection control practices advisory committee and the HICPAC/SHEA/APIC/IDSA hand hygiene task force. Infection Control and Hospital Epidemiology. 2002;**23**(S12):S3-S40

[12] Jang TH, Wu S, Kirzner D, Moore C, Youssef G, Tong A, et al. Focus group study of hand hygiene practice among healthcare workers in a teaching hospital in Toronto, Canada. Infection Control and Hospital Epidemiology. 2010;**31**(2):144-150

[13] Mazi W, Senok AC, Al-Kahldy S, Abdullah D. Implementation of the world health organization hand hygiene improvement strategy in critical care units. Antimicrobial Resistance and Infection Control. 2013;**2**(1):15

[14] Allegranzi B, Pittet D. Role of hand hygiene in healthcare-associated infection prevention. Journal of Hospital Infection. 2009;**73**(4):305-315. DOI: 10.1016/j.jhin.2009.04.019

[15] Duerink DO, Farida H, Nagelkerke NJD, Wahyono H, Keuter M, Lestari ES, et al. Preventing nosocomial infections: Improving compliance with standard precautions in an Indonesian

teaching hospital. Journal of Hospital Infection. 2006;**64**(1):36-43

[16] Ahmed QA, Memish ZA, Allegranzi B, Pittet D. Muslim health-care workers and alcohol-based handrubs. The Lancet. 2006;**367**(9515):1025-1027

[17] Liu L, Johnson H, Cousens S, Perin J, Scott S, Lawn J, et al. Global, regional, and national causes of child mortality: An updated systematic analysis for 2010 with time trends since 2000. The Lancet. 2012;**379**(9832):2151-2161. DOI: 10.1016/s0140-6736(12)60560-1

[18] Ejemot-Nwadiaro RI, Ehiri JE, Meremikwu MM, Critchley JA. Hand washing for preventing diarrhoea. Cochrane Database of Systematic Reviews. 2008;**1**:1-79

[19] Aiello A, Coulborn R, Perez V, Larson E. Effect of hand hygiene on infectious disease risk in the community setting: A meta-analysis. American Journal of Public Health. 2008;**98**(8):1372-1381. DOI: 10.2105/ajph.2007.124610

[20] Azor-Martínez E, Cobos-Carrascosa E, Gimenez-Sanchez F, Martínez-López J, Garrido-Fernández P, Santisteban-Martínez J, et al. Effectiveness of a multifactorial handwashing program to reduce school absenteeism due to acute gastroenteritis. The Pediatric Infectious Disease Journal. 2014;**33**(2):e34-e39. DOI: 10.1097/inf.0000000000000040

[21] Lau CH, Springston EE, Sohn MW, Mason I, Gadola E, Damitz M, et al. Hand hygiene instruction decreases illness-related absenteeism in elementary schools: A prospective cohort study. BMC Pediatrics. 1 Dec 2012;**12**(1):52. DOI: 10.1186/1471-2431-12-52

[22] Bowen A, Agboatwalla M, Luby S, Tobery T, Ayers T, Hoekstra R. Association between intensive handwashing promotion and child development in Karachi, Pakistan. Archives of Pediatrics & Adolescent Medicine. 2012;**166**(11):1037. DOI: 10.1001/archpediatrics.2012.1181

[23] Fasanya BK, Shofoluwe M. Occupational ergonomics: Emerging approaches toward improved worker productivity and injury reduction. In: International Conference on Applied Human Factors and Ergonomics. Cham: Springer; 2018. pp. 385-395

[24] Papastavrou JD, Lehto MR. Improving the effectiveness of warnings by increasing the appropriateness of their information content: Some hypotheses about human compliance. Safety Science. 1996;**21**(3):175-189

[25] Zimakoff J, Kjelsberg ABB, Larsen SO, Holstein B. A multicenter questionnaire investigation of attitudes toward hand hygiene, assessed by the staff in fifteen hospitals in Denmark and Norway. American Journal of Infection Control. 1992;**20**(2):58-64